NONLINEAR DYNAMICS, CHAOS AND FRACTALS

with Applications to Geological Systems

Editor: G.V. Middleton

Organizers: G.V. Middleton and B.D. Loncarevic

Authors: John Chadam, Anthony Fowler, David Goodings,

Gerard Middleton, Dan Roach, Herbert Shaw, and Donald Turcotte.

Geological Association of Canada

Short Course Notes

Volume 9

Toronto, Ontario

May 25-26, 1991

The various chapters in this volume have been prepared primarily for short course participants at the 1991 Toronto GAC Annual Meeting. Contributions have been reviewed by the editors. All comments and requests should be sent to the authors, who retain responsibility for their specific contributions.

Canadian Cataloguing in Publication Data

Main entry under title:

Nonlinear dynamics, chaos and fractals with applications to geological systems

(Short Course Notes; v. 9)
Includes bibliographic references.
ISBN 0-919216-47-1

1. Geology--Mathematics. 2. Dynamics. 3. Nonlinear theories. 4. Fractals. 5. Chaotic behavior in systems. I. Middleton, Gerard V. (Gerard Viner), 1931- . II. Geological Association of Canada. III. Series: Short course notes (Geological Association of Canada); v.9.

QE33.2.M3N66 1991 550.1'51 C91-090320-4

Geological Association of Canada
Department of Earth Sciences
Memorial University of Newfoundland
St. John's Newfoundland
Canada A1B 3X5

PREFACE

These notes are designed to introduce geologists to the new interdisciplinary body of knowledge now generally known as "Nonlinear Dynamics", or informally as "Dynamical Chaos". The meaning of these terms is sketched informally in Chapter 1. Chapters 2 to 5 are designed to give a fairly systematic introduction to the most important topics, and the other Chapters give illustrations of geological, geochemical, and geophysical applications, and develop the theory further.

All new fields of science rapidly develop their own jargon, and this is just as true of Nonlinear Dynamics as it is of other disciplines. The situation is possibly worse than usual in Nonlinear Dynamics, because many of the terms have been taken over from branches of mathematics that have been little used by scientists until recently. A Chaossary, or glossary of terms used in Nonlinear Dynamics, has been provided to help the reader. Unfortunately, the editor cannot guarantee that all the technical terms used in the Notes appear in the Chaossary -- nor can he guarantee that every term has been correctly defined -- but he does hope that the glossary will be some help to the beginner.

The organizers wishes to thank the authors for their contributions, which have cost them some labour and much harassment from the editor, borne in good humour. We particularly thank Tony Fowler and Dan Roach for providing a chapter on some of the newer techniques of time series analysis, at very short notice. The authors' work has truly been a labour of love: they do not receive stipend or royalties, and the academic profession accords little recognition to the work involved in producing such "non-refereed" publications. The editor's experience, however, is that, refereed or not, inexpensive Course Notes, are probably read more frequently and more carefully than many archival publications. Not just the editor, but the profession itself, owes a real debt of gratitude to those whose enthusiasm for their subject make them willing to undertake such a task.

Special thanks are due to Peter Russell for providing the design used on the cover.

Gerard Middleton
Bosko Loncarevic

Coorganizers

TABLE OF CONTENTS

CHAPTER 1: INTRODUCTION

CHAPTER 2: FRACTALS AND FRACTAL DIMENSION

CHAPTER 3: NONLINEAR DIFFERENTIAL EQUATIONS AND ATTRACTORS

CHAPTER 4: CHAOS IN A TIME SERIES

CHAPTER 5: ROUTES TO CHAOS

CHAPTER 1

INTRODUCTION

Gerard Middleton
Department of Geology
McMaster University
Hamilton, Ontario
L8S 4M1

> In this physical world there is no real chaos; all is in fact orderly; all is ordered by physical principles. Chaos is but unperceived order; it is a word indicating the limitations of the human mind and the paucity of observational facts. The words, "chaos", "accidental", "chance", "unpredictable", are conveniences behind which we hide our ignorance.
>
> Harlow Shapley
> Of Stars and Men (1958)

> ...well ordered predictable phenomena present just an archipelago in the ocean of Nature, which is intrinsically nonlinear and chaotic.
>
> V.I. Keilis-Boruk
> Reviews of Geophysics, v. 28, p.19 (1990)

1.2

1.1 The Breaking Wave

What is Chaos, and what does it have to do with Nonlinear Dynamics? What are Fractals? Are nonlinear dynamics, chaos and fractals concepts that geologists should know something about? Is there a revolution going on in science, comparable in scope to those produced by Plate Tectonics or Quantum Mechanics, as enthusiasts have claimed? Is it true that these studies have opened up "...an entirely new world, a new kind of mathematics, a fundamental breakthrough in the understanding of irregularities in nature" (Stewart, 1989, p.3)? Are we witnessing the rise of a "New Science of Complexity" (Pagels, 1988)? Or is this just the latest mathematico-philosophical fad, whose ultimate impact on science will perhaps be no greater than that of Catastrophe Theory, which preceded it in the 1950s and 60s?

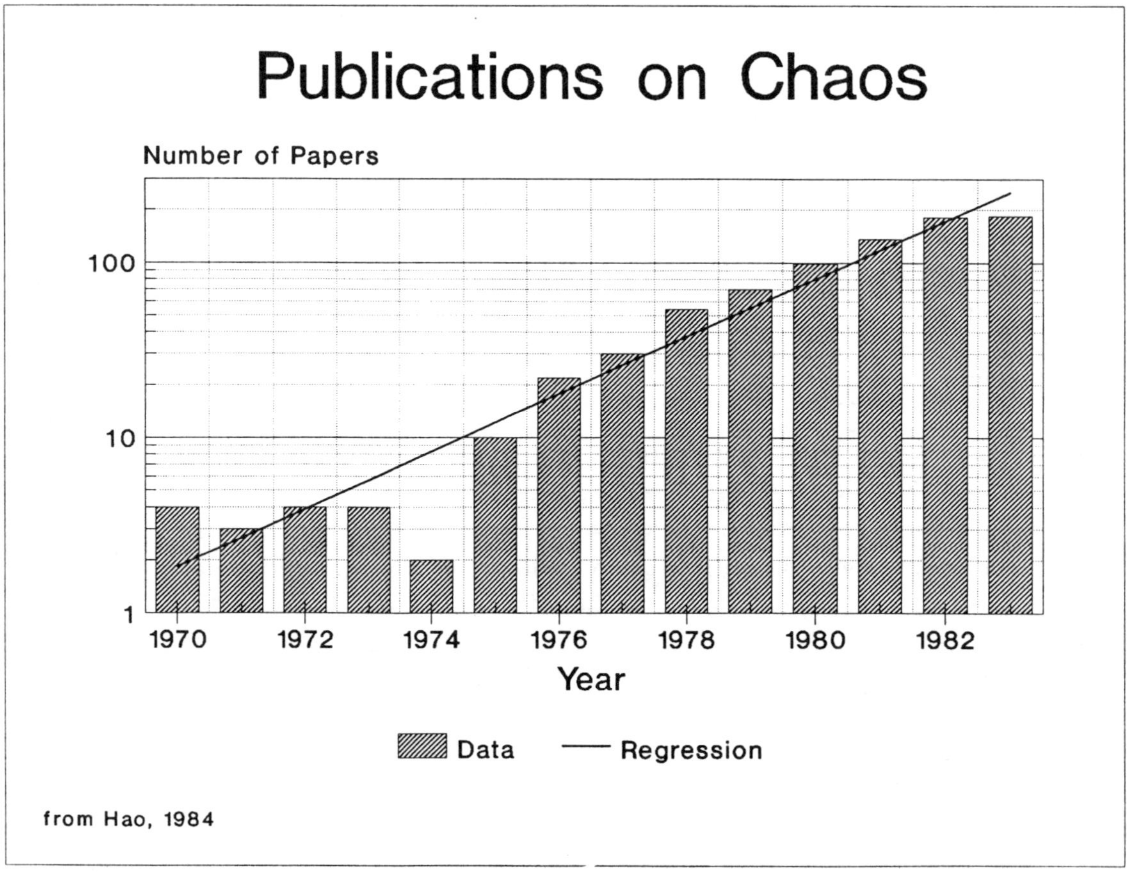

Figure 1.1: Exponential growth in number of publications on chaos. Hao (1984) gave a bibliography of 841 references, from which these data were obtained. They show that the number of papers published on chaos has doubled every two years since the mid 1970s.

One need not accept such extravagant claims to see the evidence that Nonlinear Dynamics is an emerging discipline, with its own literature (Figure 1.1), symposia, and journals (Physica D; Nonlinearity; Chaos: An Interdisciplinary Journal of Nonlinear Science). If the trend shown in Figure 1.1 is extrapolated to the 1990s, we can expect that something of the order of 1000 papers per year are already being published on chaos and related topics. My own list, which is no doubt far from complete, shows over 100 books and symposia, almost all of them published since 1980. Chaos studies are clearly recruiting the interests of an exponentially increasing number of mathematicians and scientists.

Some "breakthroughs" create a new part of science, soon more or less isolated by professional boundaries from traditional disciplines. Nonlinear Dynamics is not like that. Though originating in mathematics and physics, it has already spread rapidly through almost all other branches of science and engineering, and is beginning to make inroads in the social sciences. Its scope can be seen by consulting a recent collection of reviews published by the American Association for the Advancement of Science (Krasner, 1990). There is little doubt that it is about to extend its influence to the earth sciences: indeed, the first textbook on the subject is about to be published by one of the contributors to this Short Course (Turcotte, in press).

We hope that these notes will make it easier for geologists to understand what is going on as this new wave of theory sweeps over the earth sciences. The purpose of this brief introduction is not to attempt rigourous definitions, but simply to introduce the topics that are discussed more fully in these notes, to give some historical perspective, and to supplement the definitions given in the glossary (Chaossary: Appendix 1).

1.2 What is Nonlinear Dynamics?

Dynamics is generally understood to be that part of mechanics that deals with the relations between force and motion. The concept has been extended by theoreticians to describe the study of any systems whose state changes with time. Such systems can be described by governing equations: if the systems are continuous and deterministic, the equations will generally be sets of differential equations. Classical mechanics knows many such equations: diffusion equations, such as Fick's law in chemistry, Fourier's law of heat conduction in physics, or Darcy's law in hydrology; the wave equations; the equations of elasticity and viscous flow, and many others. Most of classical physics and geophysics has sought to obtain solutions to such sets of equations by reducing them to their simplest, linear form. Linearization is generally the only possible way to obtain analytical solutions, applicable to a wide range of initial or boundary conditions. It also makes it easier to obtain numerical solutions for special cases.

Dynamics is not only concerned with continuous systems: it may also study systems that change in discrete steps from one state to another. Such systems are described by difference equations or by maps, rather than by differential equations. Even continuous systems are frequently reduced to sets of (finite) difference equations in order to obtain numerical solutions.

Nor is dynamics confined to dealing with fully deterministic systems. Its range can be extended by incorporating "random variables" into the governing equations. Such sets of equations can then be described as stochastic models: predictions about some future state of the system are uncertain and at best it is only feasible to assign probabilities to a range of possible outcomes. In such cases, the modeller may believe, with Harlow Shapley, that the uncertain outcome is simply a result of incorporating random variables, a course of action forced by incomplete understanding of the dynamic system. In principle, at least, this type of uncertainty can be reduced as understanding improves and the model become more complete and complex.

Nonlinear dynamics is the study of systems whose governing equations are nonlinear. It seems, therefore, that it should differ from linear dynamics only quantitatively: the equations will be more complicated and harder to solve. What has made nonlinear dynamics a new science, and not just a more difficult branch of the old dynamics, has been the realization that nonlinearity causes behaviour that is not only extraordinarily complex, even for very simple sets of dynamic equations, but also that is qualitatively different from any behaviour shown by linear systems.

1.3 What is Chaos?

The dictionary definition of chaos is "total disorder or confusion" (American Heritage). The existence of natural chaotic states has long been recognized, or at least postulated, by sciences that invoke random variables or stochastic models. Statistical mechanics and statistical theories of turbulence are just two well-developed examples. The term "chaos", however, was rarely used to describe these phenomena. Its popularity as a scientific term is new and probably stems from the title of a paper by Li and Yorke (1975).

Before the development of the new ideas about chaos, highly irregular behaviour was ascribed to the action of one or more random variables. Random variables are either simply postulated to be completely random, or their average properties are thought to result from summing the action of very many separate variables, which are individually unknown and perhaps even in principle unknowable. Random behaviour of this type is essentially the product, either of inherent ignorance, or of very large numbers of degrees of freedom in the system.

In contrast, it is now known that some types of highly irregular behaviour (deterministic chaos) can result from the action of a small number of variables in a system of nonlinear governing equations. In principle, therefore, it is completely deterministic -- and the same pattern of chaos could be reproduced by the same system if the initial conditions were known with perfect accuracy. Of course, this is a very big "if" -- in reality, most nonlinear systems, even those consisting of only a few sets of equations, cannot be solved analytically. Numerical solutions must be obtained, using computers whose precision is necessarily limited. So even if the initial conditions were known perfectly, the intermediate steps in the calculation could only be carried out by numerical approximation, and thus states of the system remote from the starting state could be known only approximately.

But the real situation is worse. The main distinguishing characteristic of chaotic systems is that "errors" cumulate exponentially. Such systems display "sensitive dependence on initial conditions" (Ruelle, 1979). Linear systems, even when they incorporate stochastic variables, are not like that. We learn in elementary physics laboratories to estimate probable errors for our calculated results by considering the effects of the small errors that we are likely to make in each of our measurements. With reasonable luck, some errors "cancel out" -- but even in the worst case we assume they may simply add up. In nonlinear chaotic systems, this is not the case: very small "errors" (differences in initial conditions) soon give rise to very large differences in computed results -- and this would be true even if the results could be computed to perfect accuracy.

Sensitive dependence on initial conditions was known to Poincaré and other mathematicians by the turn of the century, but its impact on science was not felt until the work of Lorenz (1963) and Hénon and Heiles (1964). In both cases it was rediscovered "experimentally" in the process of investigating numerical solutions of sets of equations thought to be highly simplified models of natural phenomena (in meteorology and astronomy respectively).

Sensitive dependence on initial conditions does not necessarily imply that future states of chaotic systems are totally unknown. In dissipative systems (and that includes most systems of interest in the earth sciences) future states converge on (though they never quite reach) a "strange attractor" -- a geometric form (or alternatively, a set of solutions) which though it occupies only a part of the possible solution space, is called "strange" because it is not a simple point, line or surface in that space, but instead is a set of points with fractal properties. A classic example was described by Lorenz (1963), though the term itself was first used by Ruelle and Takens (1971).

1.4 What are Fractals?

Fractals are probably better known to geologists than most of the concepts discussed so far (for example, see the symposium volume edited by Scholtz and Mandelbrot, 1989). This is because geologists observe Nature, and as Mandelbrot (1982, p.1) has exhorted "Clouds are not spheres, mountains are not cones, coastlines are not circles, and bark is not smooth, nor does lightening travel in a straight line." Many natural phenomena seem to be best described, not by the old Euclidean geometry, but by a new fractal geometry. Fractals have been studied by mathematicians since the beginning of the century, though the term was introduced by Mandelbrot in the first edition of his book "Les Objets Fractals", which appeared in 1975. If an object has a complex form, such that it displays self-similarity at difference scales, and cannot be well represented by a single "characteristic length", then it displays some of the properties typical of fractals.

Fractals began as mathematical curiosities, but it has been recognized since the 1970s, that they can be used to describe many phenomena, from clouds and landscapes to the size distributions of oil fields and ore bodies, and the properties of crushed rock or pore networks. Fractals have their own inherent interest as a new form of descriptive geometry, but we do not have time in this course to explore this application of the concept. Instead, we will encounter fractals objects mainly because they (and particularly strange attractors) are naturally generated by nonlinear dynamical systems. Unexpectedly, chaos itself may have fractal properties, and nonlinear dynamics therefore opens up the possibility of explaining many fractal objects, rather than merely describing them.

It is a happy historical "accident" that nonlinear dynamics and fractal geometry developed more or less independently at about the same time. Not coincidentally, this was a time when use of computers for numerical experimentation became possible and widespread.

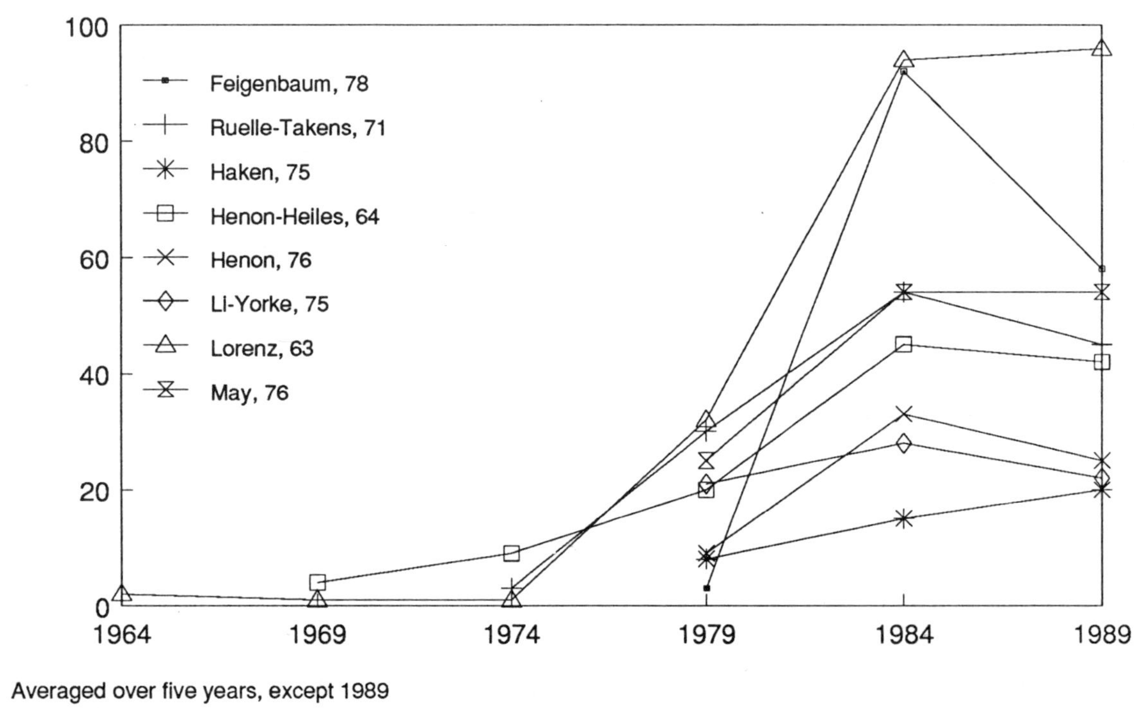

Figure 1.2: Citations per year to classic papers on Chaos. The data were taken from Science Citation Index, Five Year Summary Volumes, and averaged for the years following the publication date. The data for 1989 are the number of citations recorded in the 1989 volume.

1.5 Chaos and Plate Tectonics.

Though there may be good historical reasons why nonlinear dynamics and fractal geometry developed at about the same time, it is surely mere coincidence that this is also true of plate tectonics. The roots of both "revolutions" can be traced back to the early part of the century, and the key papers appeared in 1963 (Vine and Matthews, 1963; Lorenz, 1963) but that is almost the end to the similarities.

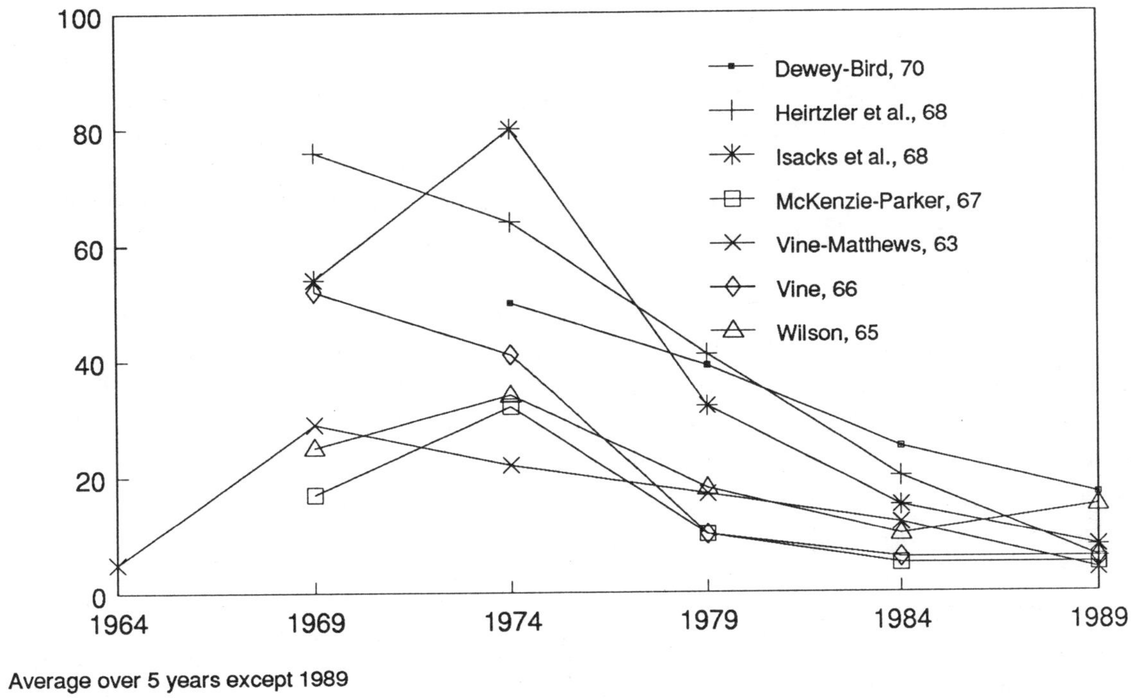

Figure 1.3: Citations per year to classic papers on Plate Tectonics, from Science Citations Index. Comparison of this figure with Figure 2.2 shows that the significance of early papers on plate tectonics was recognized very early, compared with the classics of chaos. The plate tectonic "revolution" is now an accomplished fact, whereas the chaos "revolution" is still in the making.

The Lorenz paper appeared in a relatively obscure journal, read mostly by professional meteorologists. Hénon and Heiles (1964) published in a specialist journal devoted to astrophysics. Vine and Matthews published in Nature. Though the original paper attracted relatively little attention, it was soon followed by others (Wilson, 1965; Vine, 1966; McKenzie and Parker, 1967; Morgan, 1968; Heirtzler et al., 1968; Isacks et al., 1968) that set off frenzied scientific activity -- an activity that was, however, confined to the established earth sciences. The revolution was proclaimed early (e.g., Isacks et al., 1968; Dewey and Bird, 1970). The first collection of "classic" papers and a textbook appeared only 10 years after the first paper (Cox, 1973; LePichon, Francheteau, and Bonnin, 1973).

The Lorenz (1963) and Hénon and Heiles (1964) papers remained almost unknown for 10 years. When May (1976) published his influential paper on "Simple mathematical models with very complicated dynamics" he was apparently unaware of this earlier work. The pioneer work was done in several different fields, including pure mathematics (Smale, 1967; Li and Yorke, 1975), astronomy (Hénon), meteorology (Lorenz), physics (Ruelle and Takens, 1971; Rössler, 1976; Feigenbaum, 1978), and biology (May is an ecologist trained as a mathematical physicist). It did not attract widespread attention until May's paper was published in Nature in the mid-1970s. The early papers were theoretical, and frenzied activity began only as the first papers appeared reporting well-documented examples of chaos in experimental systems (in fluid between rotating cylinders, Gollub and Swinney, 1975; Swinney and Gollub, 1979; in thermal convection, Libchaber and Maurer, 1979; and in chemical reactions, Roux et al., 1980) or using chaos theory to explain known, but previously puzzling, irregular oscillations in lasers (Haken, 1975), chemical systems (Prigogine and Lefever, 1968; Rössler, 1972) and biological systems (Mackey and Glass, 1977). By the late 1970s, however, the "revolution" was in full swing, and was the subject of many interdisciplinary symposia (e.g., New York Academy of Sciences, Gurel and Rössler, 1979; Helleman, 1980). The first texts appeared in the mid 1980s (Guckenheimer and Holmes, 1983; Bergé, Pomeau and Vidal, 1984; Schuster, 1984) together with the first collections of "classic papers" (Cvitanovic, 1984; Hao Bai-Lin, 1984 -- curiously enough, there is almost no overlap in the papers chosen by the two editors.)

Thus plate tectonics was a revolution essentially confined to one existing scientific discipline. It was rapidly recognized and developed, and within twenty years it was firmly established as part of orthodox geology, expounded in all freshman textbooks. Research papers using plate tectonic concepts no longer bother to cite the pioneer papers.

Nonlinear chaos is a revolution with multidisciplinary origins and methodological implications for all of science, engineering and the social sciences. The classic papers lay dormant for ten years. Exponential growth of research activity began only in the mid 1970s, and the topic is still "hot" and controversial in many disciplines today. The first texts and reprint volumes appeared only a few years ago, and the subject has not yet worked its way into the undergraduate curriculum in most departments (even of math or physics), let alone into freshman texts.

Welcome to the "revolution" ! Its success as a new discipline is not yet assured, but excitement is guaranteed.

1.6 Acknowledgements

I thank David Goodings for arousing and then supporting my interest in chaos. He and Bosko Loncarevic provided useful comments on an early draft of this chapter. Jim Kramer drew my attention to the quotation from Harlow Shapley.

1.7 References on Chaos

Bergé, Pierre, Y. Pomeau, and C. Vidal, 1986, Order within Chaos. Towards a deterministic approach to turbulence. N.Y., John Wiley and Sons, translated from the 1984 French edition by L. Tucherman, 329 p.

Cvitanovic, Predrag, ed., 1984, Universality in Chaos. Bristol, Adam Hilger Ltd., 511 p.

Feigenbaum, M.J., 1978, Qunatitative universality for a class of nonlinear transformations. Jour. Statistical Physics, v.19, p.25-52.

Gollub, J.P., and H.L. Swinney, 1975, Onset of turbulence in a rotating fluid. Physical Review Letters, v.35, p.927-930.

Guckenheimer, John and P. Holmes, 1983, Nonlinear Oscillations, Dynamical Systems, and Bifurcations of Vector Fields. N.Y., Springer-Verlag, 453 p.

Gurel, G., and O.E. Rössler, eds., 1979, Bifurcation Theory and Applications in Scientific Disciplines. Annales N.Y. Acad. Sci., v.316.

Haken, H., 1975, Analogy between higher instabilities in fluids and lasers. Physics Letters, v.53A, p.77-78.

Hao Bai-Lin, ed.,1984, Chaos. Singapore, World Scientific, 576 p.

Helleman, R.H.G., ed., 1980, Nonlinear Dynamics. Annals N.Y. Acad. Sci., v.357.

Hénon, M. and C. Heiles, 1964, The applicability of the third integral of motion: some numerical experiments. Astronomical Jour., v.69, p.73-79. (This paper was featured as a "citation classic" in Current Contents, Jan. 25, 1988, p.18. It has been cited over 500 times. Hénon contributed an interesting brief reminiscence about how the paper came to be written.)

Hénon, M., 1976, A two dimensional mapping with a strange attractor. Communications Math. Phys., v.50, p.69-77.

Holmes, P., 1977, 'Strange' phenomena in dynamical systems and their physical implications. Applied Math. Modelling, v.1, p.362-366.

Krasner, Saul, ed., 1990, The Ubiquity of Chaos. Washington, DC, Amer. Assoc. Advancement Sci., 247 p.

Li, Tien-Yien and James A. Yorke, 1975, Period three implies chaos. American Mathematical Monthly, v.82, p.985-992.

Libchaber, A. and J. Maurer, 1979, Une expérience de Rayleigh-Bénard de géométrie réduite; multiplication, accrochage, et démultiplication de fréquences. Jour. Physiques (Paris) Colloque 41, p.C3-C51.

Lorenz, E.N., 1963, Deterministic non-periodic flow. Jour. Atmospheric Sci., v.20, p.130-141.

Mackey, M.C., and L. Glass, 1977, Oscillation and chaos in physiological control systems. Science, v.197, p.287-289.

Mandelbrot, Benoit, 1982, The Fractal Geometry of Nature. Revised Edition, N.Y., W.H. Freeman and Co., 468 p.

May, Robert M., 1976, Simple mathematical models with very complicated dynamics. Nature, v.261, p.459-467.

Pagels, Heinz, 1988, The Dreams of Reason: The Computer and the Rise of the Sciences of Complexity. N.Y., Simon and Schuster, 352 p.

Prigogine, I., and R. Lefever, 1968, Symmetry breaking instabilities in dissipative systems, II. Jour. Chemical Physics, v.48 , p.1695-1700.

Rössler, O.E., 1972, A principle for chemical multivibration. Jour. Theoretical Biology, v.36, p.413-417.

Rössler, O.E., 1976, An equation for continuous chaos. Physics Letters, v.57A, p.397.

Roux, J.C., A. Rossi, S. Bachelart, and C. Vidal, 1980, Representation of a strange attractor from an experimental study of chemical turbulence. Physics Letters, v.77A, p.391-393. (Period doubling in the Belousof-Zhabotinski reaction)

Ruelle, D. and F. Takens, 1971, On the nature of turbulence. Communications Math. Phys., v.20, p.167-192.

Ruelle, D., 1979, Sensitive dependence on initial conditions and turbulent behavior of dynamical systems. Annals N.Y. Acad. Sci., v.316, p.408-416.

Scholtz, Christopher H., and Benoit B. Mandelbrot, eds., 1989, Fractals in Geophysics. Boston, Birkhäuser Verlag, 313 p. Originally published in Pure and Applied Geophysics, v. 131, no. 1/2.

Schuster, H.G., 1984, Deterministic Chaos: an Introduction. VCH Publishers, Inc. Second Edition, 1988.

Smale, S., 1967, Differentiable dynamical systems. Bull. Amer. Mathematical Soc., v.73, p.747-817 (Stretching and folding in strange attractors)

Stewart, Ian, 1989, Does God Play Dice? The Mathematics of Chaos. Oxford, Basil Blackwell, 317 p.

Swinney, H.L., and J.P. Gollub, 1978, The transition to turbulence. Physics Today, v.31, no. 8, p.41

Turcotte, Donald L., in press, Fractals and Chaos in Geology and Geophysics. Cambridge, Cambridge University Press.

1.8 References on Plate Tectonics

Cox, Allan, ed., 1973, Plate Tectonics and Geomagnetic Reversals. San Francisco, W.H. Freeman, 702 p.

Dewey, John F., and J.M. Bird, 1970, Mountain belts and the new global tectonics. Jour. Geophysical Research, v.75, p.2625-2647.

Heirtzler, J.R., G.O. Dickson, E.M. Herron, W.C. Pitman III, and X. Le Pichon, 1968, Marine magnetic anomalies, geomagnetic field reversals and motion of the ocean floor and continents. Jour. Geophysical Research, v.73, p.2119-2136.

Isacks, B.L., J. Oliver and L.R. Sykes, 1968, Seismology and the new global tectonics. Jour. Geophysical Research, v.73, p.5855-5699.

Le Pichon, X., J. Francheteau and J. Bonnin, 1973, Plate Tectonics. N.Y., Elsevier, 302 p.

McKenzie, D.P., and Robert Parker, 1967, The North Pacific: An example of tectonics on a sphere. Nature, v.216, p.1276-1280.

Morgan, W. Jason, 1968, Rises, trenches, great faults and crustal blocks. Jour. Geophysical Research, v.73, p.1959-1982.

Vine, F.J., 1966, Spreading of the ocean floor: new evidence. Science, v.154, p.1405-1415.

Vine, F.J., and D. Matthews, 1973, Magnetic anomalies over ocean ridges. Nature, v.199, p.947-949.

Wilson, J. Tuzo, 1965, A new class of faults and their bearing on continental drift. Nature, v.207, p.343-347.

CHAPTER 2

FRACTALS AND FRACTAL DIMENSION

David Goodings and Gerard Middleton
Departments of Physics and Geology
McMaster University
Hamilton, Ontario
L8S 4M1

2.1 Introduction

In this first of four introductory chapters on nonlinear dynamics and chaos in dissipative systems, we begin by introducing the idea of a fractal. After describing examples of naturally occurring fractal objects such as coastlines and turbulent fluids we consider a simple mathematical model of a fractal. This leads to the definition of the most important quantitative feature of a fractal object, the fractal dimension. The definition is illustrated by explicit calculations for some simple mathematical models.

2.2 Eddies and coastlines

Many natural objects have the property of looking much the same when enlarged (as in a photographic enlargement) or reduced. This feature of 'self-similarity under changes of scale' is displayed by turbulent flow in which eddies are found with sizes ranging from large-scale rotations of the fluid down to tiny vortices which ultimately dissipate much of the fluid's kinetic energy. The word 'fractal' was coined by B.Mandelbrot (1975, 1983) as a generic term for objects that are self-similar over a wide range of length scales.

The classic example of a fractal object is a coastline. (See Mandelbrot's chapter entitled, 'How long is the coast of Britain?') An ordinary map which ignores many details of a coastline can be used to provide a first estimate of its length. However, a better map exhibiting the capes and bays on an enlarged scale will yield a length for the coastline that is considerably larger than the first estimate. If δ is a length

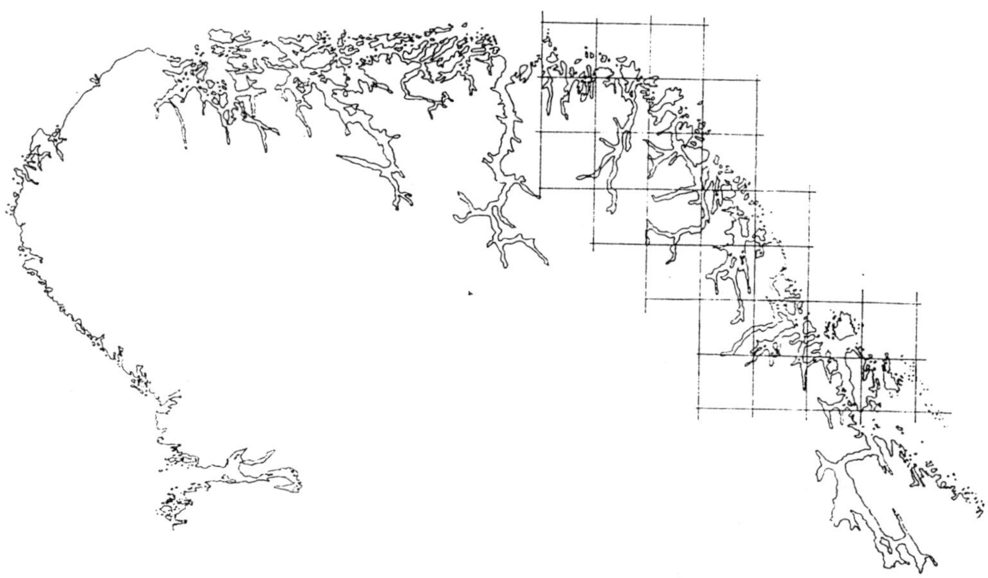

Figure 2.1 The coastline of the southern part of Norway. The outline was traced from an atlas and digitized at about 1800×1200 pixels. The square grid indicated has a spacing of $\delta \sim 50$km. Reproduced from Feder (1988) with permission.

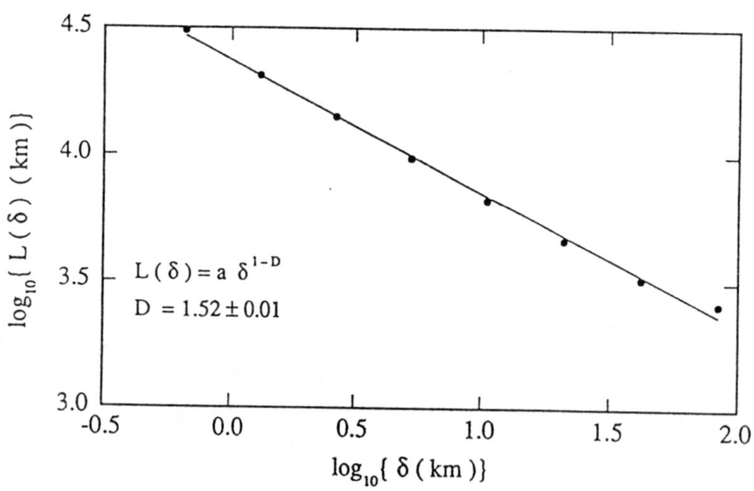

Figure 2.2 The measured length of the coastline shown in Figure 2.1 as a function of the size δ of the $\delta \times \delta$ squares used to cover the coastline on the map. The straight line in this log-log plot corresponds to the relation $L(\delta) = a\delta^{1-D}$ with $D \simeq 1.52$.

characterizing the size of the smallest features revealed on a map of a given coastline, it is found empirically that the length L depends on δ according to a relationship of the form,

$$L(\delta) \approx \frac{\text{const}}{\delta^{D-1}}, \tag{2.1}$$

where D is a constant greater than unity. The value of D depends on the particular coastline being considered, and different parts of a coastline may well exhibit different values of D. For $D > 1$ this relation clearly shows the property that the length L increases as the size of the measuring stick δ decreases. For physical objects (in contrast to mathematical models, to be described below) there is always a smallest value of δ below which the measurement process is poorly defined. Ignoring such practical matters as tides and knowing at what point to jump across the mouth of a river, one can see that in measuring the length of a real coastline, the lower limit for δ cannot be less than about one metre.

The exponent D in (2.1) is called by Mandelbrot the fractal dimension. Mandelbrot (1983) shows plots of measurements of $\log L$ against $\log\delta$ carried out for several different coastlines and national boundaries by Lewis F. Richardson (1961). A typical value for D is about 1.25.

The coastline of Norway, part of which is shown in Figure 2.1, is quite different from Richardson's coastlines and frontiers because of its numerous fjords, many of which display branchings into smaller fjords. The length of the coastline shown in Figure 2.1 has been measured by Feder (1988) as a function of δ for δ ranging from 100 km down to less than 1 km. His plot of $\log_{10} L(\delta)$ against $\log_{10}\delta$ is shown in Figure 2.2. From the slope of the straight line fitted to the measurements he deduced that $D \simeq 1.52$, a value which reflects the much more irregular nature of this coastline compared to the ones considered by Richardson.

Fractal objects in nature, such as coastlines and the sequence of branchings of a large tree, never appear precisely the same on expanding or reducing the scale on which the object is viewed. However, some mathematical models are exactly self-similar when the length scale is changed by a constant factor. Studying such models gives some insight into the mathematical description of fractals.

2.3 The Koch snowflake

Starting from an equilateral triangle each side of which has unit length, one can divide each side into three equal parts and erect an equilateral triangle on the middle segment of each side, as shown in Figure 2.3. The length L of the perimeter of the original triangle increases by a factor of 4/3 as a result of this process. If each side of the new figure is then divided into three equal parts and equilateral triangles are erected on the middle segments, the length of the perimeter again increases by a factor of 4/3. When this construction is repeated many times, the length of the perimeter after the pth stage is,

$$L = 3\left(\frac{4}{3}\right)^p, \qquad p = 0, 1, 2, \ldots \tag{2.2}$$

To cast this into the form of equation (2.1), we choose δ at the pth stage to be $(1/3)^p$ and the constant in (2.1) to be 3 corresponding to the length of the perimeter of

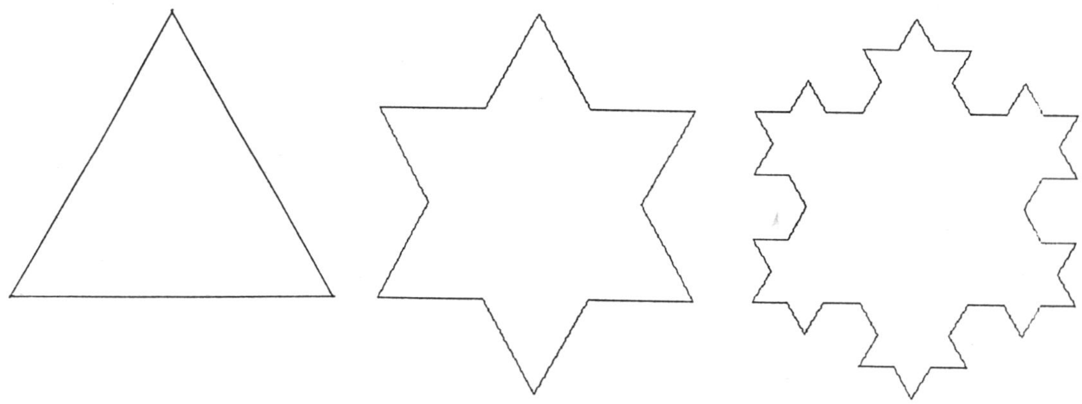

Figure 2.3 The first three stages of construction of the Koch snowflake.

the original triangle. The fractal dimension D must satisfy,

$$3\left(\frac{4}{3}\right)^p = \frac{3}{[(1/3)^p]^{D-1}}, \qquad p = 0, 1, 2, \ldots \tag{2.3}$$

from which it follows after taking logarithms (to any base) of both sides that,

$$D = \frac{\log 4}{\log 3} = 1.26186... \tag{2.4}$$

The limiting form that results after an infinite number of repetitions of the construction process is known as the Koch snowflake (von Koch, 1904). Its perimeter is a continuous curve of infinite length. (As a curiosity it may be noted that at no point on this curve does the derivative exist.) By analogy with a coastline we have found its fractal dimension to be $D = \log 4/\log 3$.

2.4 The fractal dimension by counting boxes

The fractal dimension D introduced in equation (2.1) can be generalized to handle fractal objects consisting of areas and volumes as well as lengths. However, it is more useful to introduce a general way of defining D, closely related to what mathematicians call the Hausdorff dimension.

Any fractal object is a geometrical form existing in a space of one, two or three dimensions. (Clearly, the Koch snowflake exists in a space of two or more dimensions.) Whatever the minimum spatial dimension required to portray the fractal object, imagine covering the entire object with 'boxes' of side δ. In one dimension a box is a line segment of length δ; in two dimensions it is a square of area δ^2; in three dimensions it is a cube of volume δ^3. If for given δ the number of such 'boxes' needed to cover the entire fractal object is $N(\delta)$, then the fractal dimension D of the object is given approximately by,

$$N(\delta) \approx \frac{K}{\delta^D}, \qquad \delta \text{ small} \tag{2.5}$$

where K is a constant. Taking logarithms we obtain,

$$\log N(\delta) \approx \log K + D \log(1/\delta). \tag{2.6}$$

As δ approaches zero, both $\log N(\delta)$ and $\log(1/\delta)$ become very large compared to $\log K$. In the limit $\delta \to 0$ we obtain the following precise definition for the fractal dimension D:

$$D = \lim_{\delta \to 0} \frac{\log N(\delta)}{\log(1/\delta)}. \tag{2.7}$$

As a simple example let us apply this definition to find the dimension of a nonfractal object, namely, the set of all points inside a square of unit length on each side. Clearly a single box with $\delta = 1$ covers the whole square. When $\delta = 1/2$, $2^2 = 4$ boxes are required to cover the unit square, and when $\delta = 1/2^p$, 2^{2p} boxes are needed. Applying (2.7) at the pth stage, and noting that the limit $\delta \to 0$ corresponds to $p \to \infty$, we obtain,

$$D = \lim_{p \to \infty} \frac{\log 2^{2p}}{\log 2^p} = 2. \tag{2.8}$$

It is satisfying that this simple geometrical form has a fractal dimension in agreement with the familiar concept of dimension.

2.5 Examples of calculating the fractal dimension

The definition of fractal dimension D in equation (2.7) is appealing because it can be used to estimate D for any object, irregular or regular (in a geometrical sense), fractal or nonfractal. One simply sets down a grid of boxes of size δ, counts the number $N(\delta)$ required to cover the object, repeats the process for successively smaller values of δ and attempts to extrapolate the right-hand side of (2.7) to $\delta = 0$.

Let us use this procedure to obtain the fractal dimension of the Koch snowflake of section 2.3. The required boxes are, of course, squares of side δ in this case. A single box with side $\delta = 1$ covers the entire equilateral triangle. At the first stage of erecting triangles on the middle segments of each side one requires $3 \times 3 = 9$ boxes with $\delta = 1/3$. At the second stage, however, with $\delta = 1/3^2$, one can see from Figure 2.3 that the number of boxes required is $3 \times 4 \times 3$. At the pth stage one finds that,

$$\text{for } \delta = 1/3^p, \quad N(\delta) = 3 \times 4^{p-1} \times 3 = (9/4) \times 4^p. \tag{2.9}$$

From (2.7) the fractal dimension of the Koch snowflake is,

$$D = \lim_{\delta \to 0} \frac{\log N(\delta)}{\log(1/\delta)} = \lim_{p \to \infty} \frac{\log(9/4) \times 4^p}{\log 3^p}$$

$$= \lim_{p\to\infty} \frac{\log(9/4)}{p\log 3} + \lim_{p\to\infty} \frac{p\log 4}{p\log 3} = \frac{\log 4}{\log 3}. \tag{2.10}$$

This is the same result as was obtained earlier by considering the length of the perimeter of the snowflake. Note that the limit $p \to \infty$ removes the first term of the last line of (2.10).

Figure 2.4 The first three stages of construction of the Cantor set.

As a second example let us calculate the fractal dimension of the interesting set of points first studied by G.Cantor (1883) and known as the Cantor set. The set is generated by starting with a straight line of unit length, removing the middle third of the line but leaving behind the end points, removing the middle segments of the two remaining lines but leaving behind the end points, and repeating this process an infinite number of times (see Figure 2.4). In the limit one is left with a set consisting of an infinite number of points while the total length of the undeleted line segments has shrunk to zero. What is the fractal dimension of this set of points?

Referring to Figure 2.4 one can easily see that if at the pth stage one chooses $\delta = 1/3^p$, the number of one-dimensional boxes of length δ required to cover the remaining line segments (including their end points) is 2^p. Hence, from (2.7) the fractal dimension of the Cantor set is,

$$D = \lim_{\delta\to 0} \frac{\log N(\delta)}{\log(1/\delta)} = \lim_{p\to\infty} \frac{\log 2^p}{\log 3^p} = \frac{\log 2}{\log 3}. \tag{2.11}$$

Figure 2.5 The first two stages of construction of the Sierpinski carpet.

As a final example we consider the object known as the Sierpinski carpet (Sierpinski, 1916). The first two stages of construction are shown in Figure 2.5. It may be seen that at each stage the area is reduced by a factor of 8/9. If each side of the initial square has unit length, at the first stage we require 8 'boxes' of side $\delta = 1/3$ to cover the object. At the second stage we require 8^2 'boxes' of side $\delta = 1/3^2$ to cover the object, and at the pth stage of construction we need 8^p 'boxes' of side $1/3^p$. From the box-counting formula (2.7) it follows that the fractal dimension of the Sierpinski carpet is,

$$D = \lim_{\delta \to 0} \frac{\log N(\delta)}{\log \delta} = \lim_{p \to \infty} \frac{\log \delta^p}{\log 3^p} = \frac{\log 8}{\log 3} = 1.892789\ldots \qquad (2.12)$$

The fractal dimension is less than 2 because an infinite number of repetitions of cutting out the middle squares causes the area to shrink to zero.

2.6 Fractals in geology

For geology, as for the other natural sciences, one of the attractions of fractal geometry is that it provides a system for describing complex natural objects, which can be approximated only with great difficulty using ordinary Euclidean geometry.

A common technique, used to determine whether or not a particular phenomenon is fractal, is to examine how the frequency of such phenomena changes as the scale is varied. If the frequency is directly proportional to a power of the scale, then we can claim that there is no natural or 'characteristic' scale describing the phenomenon, and that it shows at least some fractal properties.

The use of a fractal dimension to describe geomorphic features is a case in point. There is no unique 'length' of a coastline because the measured length depends on the scale of measurement. Similarly it has been claimed that there is no characteristic size (i.e. no meaningful 'mean size') of any of the following phenomena: rock pores, caves, the fragments in crushed rock, meander loops, coral reefs, eddies in the ocean, clouds, rainfall, floods, earthquakes, stratigraphic hiatuses, oil fields, or ore bodies. For example, Gutenberg and Richter established that, in a given time, the number of earthquakes N with a magnitude greater than m is given by,

$$\log N = -bm + a, \qquad (2.13)$$

where a and b are constants. It can be shown that this is equivalent to a fractal distribution relating the size of earthquakes (as measured by the area of rupture) to their frequency (Turcotte, 1991). The fractal dimension is $2b$, and generally has a value of about 1.8.

The evidence for these claims is discussed in a number of review papers (Bradbury and Reichelt, 1983; Burrough, 1984; Church and Mark, 1980; Curl, 1986; Goodchild and Mark, 1987; Plotnick, 1986; Thompson *et al.*, 1987; Turcotte, 1988, 1991). Besides Mandelbrot's classic work, there are now a number of other books that further develop the theory and give many other examples (Feder, 1988; Kaye, 1989; Peitgen and Saupe, 1988; Scholtz and Mandelbrot, 1989; Turcotte, in press).

Of course, simple mathematical models rarely provide a perfect fit for the real world, and this is true also for fractal models. Generally, size distributions are fractal over only a limited range of sizes. At very small sizes, the distribution is limited by crystalline or molecular phenomena: for example, the smallest features of pores may be limited by a minimum size of crystals which can seed on the surface of a pore. At the largest sizes, the possible size may be limited by the size of the earth, or by the thickness or strength of the rigid lithosphere, etc.

Not all fractal phenomena are adequately described by the simple models discussed in earlier parts of this chapter. There is no time in this course to discuss the ways in which the simple theory can be elaborated, but in the next section we consider briefly the possibility that some phenomena may be best described by a spectrum of fractal dimensions, rather than by a single dimension.

2.7 Multifractals

Let us return to the Cantor set discussed in Section 2.5. Note that this model (the 'uniform triadic Cantor set') can easily be elaborated by removing some other fraction than one third of each line segment, or by leaving behind segments of unequal length (say l_1 and l_2). The process can also be made nondeterministic by selecting the position and size of the segment removed at random.

Suppose, however, that we retain the deterministic character, but associate two probabilities p_1 and p_2 with each line segment. Let the initial lengths be 1/4 and 2/5, and the associated probabilities be 3/5 and 2/5 respectively. At the second stage the lengths are:

$$(1/4)(1/4) \qquad (1/4)(2/5) \qquad (2/5)(1/4) \qquad (2/5)(2/5)$$

and the associated probabilities are:

$$(3/5)(3/5) \qquad (3/5)(2/5) \qquad (2/5)(3/5) \qquad (2/5)(2/5)$$

The probabilities constitute a *measure* which is associated with a fractal *support.* The measure is not distributed at random over the support but displays a spatial correlation. This algorithm, called a *multiplicative process* (Feder, 1988, p.71) might well be a simple model for a physical process, for example, for the way in which energy is transferred from larger to smaller turbulent eddies. It can be shown (for details and further references see Feder, 1988; Lovejoy and Schertzer, 1990; Mandelbrot in Scholtz and Mandelbrot, 1989) that the process gives rise to distributions (of probability, in the Cantor set example) which cannot be adequately characterized by a single dimension. They can, however, be characterized by a spectrum of dimensions; hence the name *multifractal.*

The importance of multifractals for the earth sciences stems directly from the nature of multiplicative processes. In the earth, many of the processes generating geological phenomena may be expected to be stochastic, yet strongly correlated in space. It is not surprising, therefore, that recent studies are bringing forward strong evidence that several of the phenomena listed in the previous section are multifractal rather than fractal. Examples include strong turbulence in fluids, earthquakes, and some aspects of porosity.

We are interested in fractals partly because it has been shown that many nonlinear dynamical models display chaotic régimes that naturally give rise to fractal objects such as strange attractors. At least some of these objects have now been shown to be multifractal rather than fractal.

References

Bradbury, R.H., and Reichelt, R.E., 1983, Fractal dimension of a coral reef at ecological scales: Marine Ecology Progress Series, v.10, p.169-171.

Burrough, P.A., 1984, The application of fractal ideas to geophysical phenomena: Institute of Mathematics and its Applications, v.20, p.36-42.

Cantor, G., 1883, Grundlagen einer allgemeinen Mannigfaltigkeitslehre: Mathematische Annalen, v.21, p.545-591.

Church, M., and Mark, D.M., 1980, On size and scale in geomorphology: Progress in Physical Geography, v.4, p.342-390.

Curl, R.L. 1986, Fractal dimensions and geometries of caves: Mathematical Geology, v.18, p.765-784.

Feder, J., 1988, Fractals, Plenum, New York.

Goodchild, M.F., and Mark, D.M., 1987, The fractal nature of geographic phenomena: Annals Assoc. Amer. Geographers, v.77, p.265-278.

Kaye, B.H., 1989, A random walk through fractal dimensions, VCH Publishers, New York.

Lovejoy, S., and Schertzer, D., 1990, Our multifractal atmosphere: a unique laboratory for nonlinear dynamics: Physics in Canada, July 1990, p.62-71.

Mandelbrot, B.B., 1975, Les objets fractals: forme, hazard et dimension, Flammerion, Paris.

Mandelbrot, B.B., 1983, The fractal geometry of nature, Freeman, San Francisco.

Peitgen, H.-O., and Saupe,D., eds., 1988, The science of fractal images, Springer-Verlag, New York.

Plotnick, R.E., 1986, A fractal model for the distribution of stratigraphic hiatuses: Jour. Geology, v.94, p.885-890.

Richardson, L.F., 1961, The problem of contiguity: an appendix of statistics of deadly quarrels: General Systems Yearbook, v.6, p.139-187.

Scholtz, C.H., and Mandelbrot, B.B., eds., 1989, Fractals in geophysics, Birkhäuser Verlag, Boston. Originally published in Applied Geophysics, v.131, no.1/2.

Sierpinski, W., 1916, Sur une courbe cantorienne qui contient une image biunivoque et continue de toute courbe donnée: Comptes Rendus (Paris), v.162, p.629.

Thompson, A.H., Katz, A.J., and Krohn, C.E., 1987, The microgeometry and transport properties of sedimentary rock: Advances in Physics, v.36, p.625-694.

Turcotte, D.L., 1988, Fractals in fluid mechanics: Ann. Rev. Fluid Mech., v.20, p.5-16.

2.10

Turcotte, D.L., 1991, Fractals in geology: What are they and what are they good for? GSA Today, v.1, no.1, p.1-4.

Turcotte, D.L., Fractals and chaos in geology and geophysics, Cambridge UP, Cambridge, U.K.

von Koch, H., 1904, Sur une courbe sans tangente, obtenue par une construction géométrique élémentaire: Arkiv för Matematik, Astronomi och Fysik, v.1, p.681-704.

CHAPTER 3

NONLINEAR DIFFERENTIAL EQUATIONS AND ATTRACTORS

David Goodings
Physics Department
McMaster University
Hamilton, Ontario
L8S 4M1

3.1 Introduction

In this chapter we consider a variety of systems described by nonlinear differential equations. The idea of an attractor is introduced and illustrated by means of the damped harmonic oscillator and the van der Pol equation. An attractor having a much more complex structure arises from the set of differential equations introduced by E.N. Lorenz to provide a simple model of Rayleigh-Bénard convection. Known as a strange or chaotic attractor, the two-lobed Lorenz attractor consists of an infinite number of sheets and has a fractal dimension of 2.06. The concept of Lyapunov exponents is introduced and shown to provide a mathematically precise way of defining chaotic behaviour.

3.2 Example of a point attractor

In order to introduce the concept of an attractor, let us consider a damped harmonic oscillator described by the linear differential equation,

$$\frac{d^2\theta}{dt^2} + \gamma\frac{d\theta}{dt} + \omega^2\theta = 0. \tag{3.1}$$

This equation is linear because each term contains only one of $\theta(t)$ or one of its time derivatives to the first power. It is a good approximation to the equation of motion of a simple pendulum with damping, provided $\gamma > 0$ and the amplitude of oscillation is small. In this case θ is the angle between the pendulum and the vertical direction, ω is the angular frequency of the undamped oscillator (corresponding to $\gamma = 0$) and γ is the damping constant. If γ is small compared to ω, the oscillator will make several oscillations before effectively coming to rest at $\theta = 0$.

A second-order linear differential equation such as (3.1) can always be expressed as two first-order linear differential equations:

$$\begin{aligned} \frac{d\theta}{dt} &= u, \\ \frac{du}{dt} &= -\gamma u - \omega^2 \theta. \end{aligned} \tag{3.2}$$

The differential equations considered in this chapter will always be put in this form. The two-dimensional space consisting of the variables θ and u will be called the *state space*. (The term *phase space* is often casually employed by writers on chaos, much to the chagrin of physicists versed in analytical mechanics and statistical mechanics for whom the term is used only to denote a $2n$-dimensional space consisting of n coordinates and their corresponding momenta.)

Although equations (3.2) can be solved in closed form, we are not particularly interested here in the mathematical expression for the solution. The nature of the solutions is revealed by Figure 3.1 which shows plots of $u = d\theta/dt$ against θ for three cases. When $\gamma > 0$ the solution to (3.2) is a spiral curve which starts from some point representing the initial values of θ and u and winds into the origin. Solutions starting from different initial values of θ and u will look very much the same. In every case, when $\gamma > 0$, the solution approaches the origin implying that eventually the pendulum will hang vertically ($\theta = 0$) and remain motionless ($u = d\theta/dt = 0$). The origin is said to be a *point attractor* since that is where the solutions end up at large times.

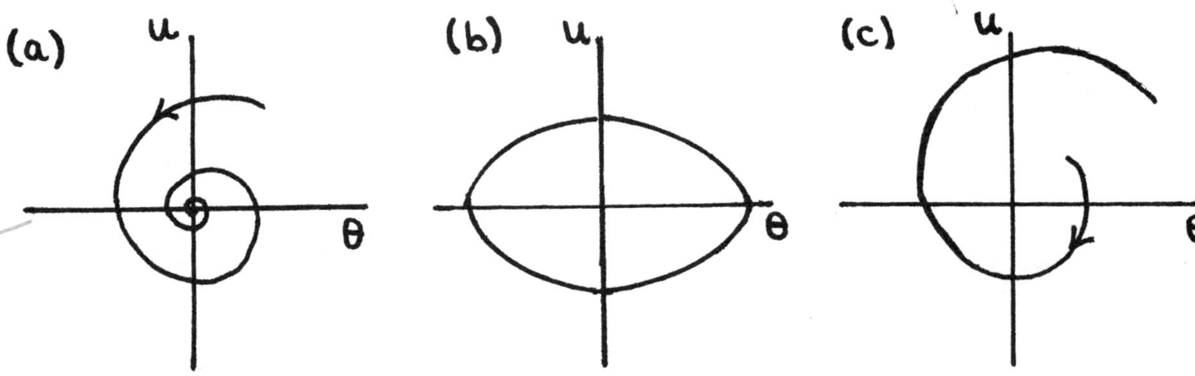

Figure 3.1 Typical trajectories of the damped harmonic oscillator for three cases: (a) $\gamma > 0$ corresponding to damping; (b) $\gamma = 0$ corresponding to no damping; (c) $\gamma < 0$ corresponding to 'antidamping'. In case (a) the origin is a point attractor.

When $\gamma = 0$ equations (3.2) describe the motion of an undamped harmonic oscillator. The solution in this case, starting from given values of θ and u, is an ellipse as shown in Figure 3.1(b). The trajectory (meaning the path followed by the variables θ and u in the state space) is closed and repeats indefinitely. Since the ellipse is not approached with increasing time from some other part of the state space, it is not regarded as an attractor.

To complete the list of possible mathematical solutions of equations (3.2), Figure 3.1(c) shows a typical trajectory when $\gamma < 0$. The solution spirals outward, no matter what the initial values of θ and $d\theta/dt$ are. While this case of 'antidamping' is unphysical in the present context since it requires a constant flow of energy into the system, it will be found to arise in the case of the van der Pol oscillator described in the next section.

The principle result of this section may be restated. For the physically meaningful case of a damped harmonic oscillator ($\gamma > 0$), all trajectories, regardless of the initial values of θ and $d\theta/dt$, spiral in toward the origin which is a point attractor.

3.3 Example of a limit cycle

A second type of attractor is exhibited by the nonlinear differential equation known as the van der Pol equation (van der Pol, 1927):

$$\frac{d^2\theta}{dt^2} + \gamma_0 \left(\frac{\theta^2 - \theta_0^2}{\theta_0^2} \right) \frac{d\theta}{dt} + \omega^2 \theta = 0. \tag{3.3}$$

Here γ_0 and θ_0 are both positive constants. This can be broken up into two first-order differential equations,

$$\begin{aligned} \frac{d\theta}{dt} &= u, \\ \frac{du}{dt} &= -\gamma_0 \left(\frac{\theta^2 - \theta_0^2}{\theta_0^2} \right) u - \omega^2 \theta, \end{aligned} \tag{3.4}$$

which are clearly nonlinear because of the term involving $\theta^2 u$ on the right-hand side of the second equation.

Comparing (3.4) with (3.2) we see that the damping constant γ has been replaced by a damping function,

$$\gamma(\theta) = \gamma_0 \left(\frac{\theta^2 - \theta_0^2}{\theta_0^2} \right). \tag{3.5}$$

This has the property that when $|\theta| < \theta_0$, $\gamma(\theta)$ is negative and the motion is antidamped. Thus, if the oscillator starts out with oscillations of small amplitude, the amplitude will grow (requiring an input of energy) until it reaches approximately θ_0. It will then continue thereafter to oscillate with approximately that amplitude. On the other hand, if the oscillator starts out making oscillations of large amplitude, the amplitude will steadily decrease (as a result of energy dissipation) until it reaches approximately θ_0, remaining at that amplitude thereafter. Typical trajectories corresponding to initial amplitudes less than and greater than θ_0 are shown in Figure 3.2. Both trajectories may be seen to approach a slightly distorted elliptical curve which is called a *limit cycle*. For given values of γ_0 and θ_0 the limit cycle is uniquely determined. It is the attractor associated with the van der Pol equation since all trajectories are attracted toward it.

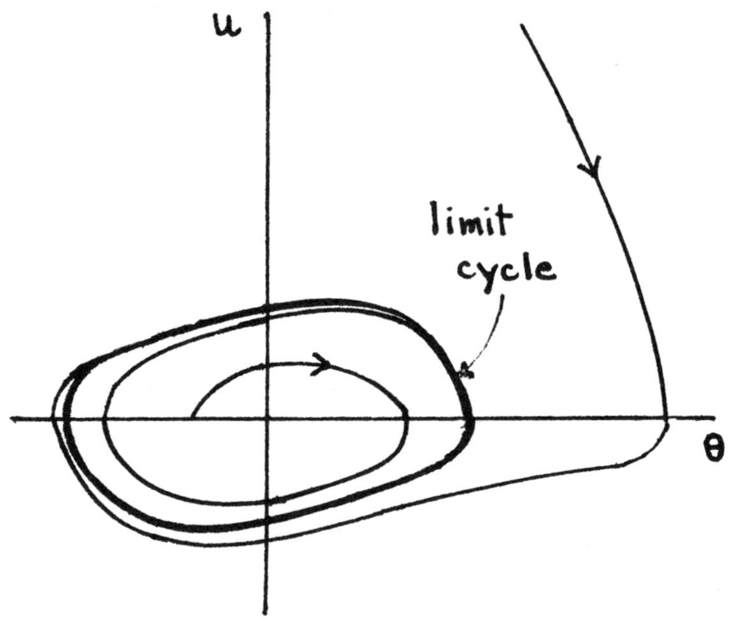

Figure 3.2 Typical trajectories of the van der Pol oscillator. Both types of trajectories approach the same curve which is an attractor known as a limit cycle.

In general an attractor associated with a set of nonlinear differential equations is a geometrical form which characterizes the long-time behaviour of the trajectories representing the solutions. Usually the trajectories approach the attractor asymptotically. This happens in the present example of the limit cycle of the van der Pol equation. The point attractor of the damped harmonic oscillator is also approached asymptotically. In practice, however, after a certain amount of time has elapsed, the trajectories are so close to the attractor that the difference cannot be distinguished on plots like Figures 3.1(a) and 3.2. During the time it takes for a trajectory to get close to the attractor, the behaviour of the system is said to undergo an *initial transient*.

3.4 The torus attractor

For the sake of completeness a third type of attractor may be briefly mentioned at this point, namely a *torus*. A system described by a set of nonlinear differential equations may, after an initial transient, develop a periodic or almost periodic motion involving *two* frequencies. A good example occurs in the configuration known as Couette flow in which fluid occupies the region between two vertical cylinders.

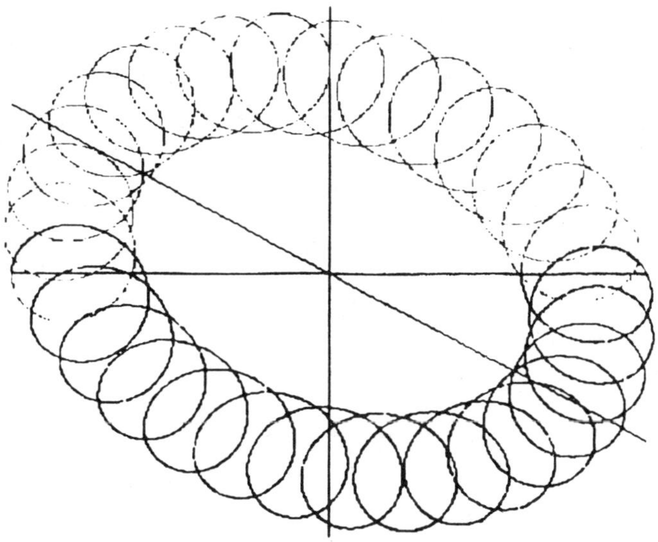

Figure 3.3 A trajectory on a torus. The initial transient is not shown.

With the outer cylinder stationary the inner cylinder is made to rotate at an angular velocity Ω. When Ω exceeds a critical value Ω_c, the fluid displays a set of vortices extending toroidally all the way around the inner cylinder, like a stack of doughnuts on a vertical peg. The fluid circulates in each vortex at a frequency f_1. However, if Ω is further increased, the vortices become wavy in the azimuthal direction, and the fluid circulates around the inner cylinder at a frequency f_2 which is much less than f_1. After suitable mathematical transformations the motion of such a system can be portrayed as a trajectory winding around a torus (in fact, approaching the torus asymptotically), as in Figure 3.3. One frequency of the motion is associated with the trajectory moving around the larger circular cross-section of the torus, the other moving around the smaller circular cross-section. The actual trajectory is the combination of both circular motions and has a helical form. If the two frequencies are commensurate (meaning that $f_1/f_2 = m/n$ where m and n are integers), the trajectory will be a closed curve and the motion will be periodic. However, if the two frequencies are incommensurate, the trajectory will be a curve which winds forever on the surface of the torus without closing on itself. Since the motion may be nearly periodic but not exactly so, it is said to be *quasiperiodic*. In both cases the attractor is the torus.

3.5 Rayleigh-Bénard convection and the Lorenz equations

The attractors encountered so far have simple geometrical forms. Some attractors, however, are not at all simple. In this section the background leading to the Lorenz equations is described. For certain ranges of the parameters in these equations the trajectories representing the solutions approach a strange geometrical form called, appropriately, a *strange attractor*. Its geometry will be described in the following section.

Consider a fluid between two infinite horizontal conducting plates, the upper plate maintained at temperature T_0 and the lower plate at a higher temperature $T_0 + \Delta T$. Provided the separation d between the plates is not large and provided ΔT is not large, heat will be transported from the lower plate to the upper plate by ordinary heat conduction involving the diffusion of energy by the molecules of the fluid. There is no motion of the fluid on a macroscopic scale. However, if the plates are brought close together or if the temperature difference between the plates is sufficiently large, the fluid will spontaneously develop convective motion in cells of the type shown in Figure 3.4. This is the configuration known as Rayleigh-Bénard convection.

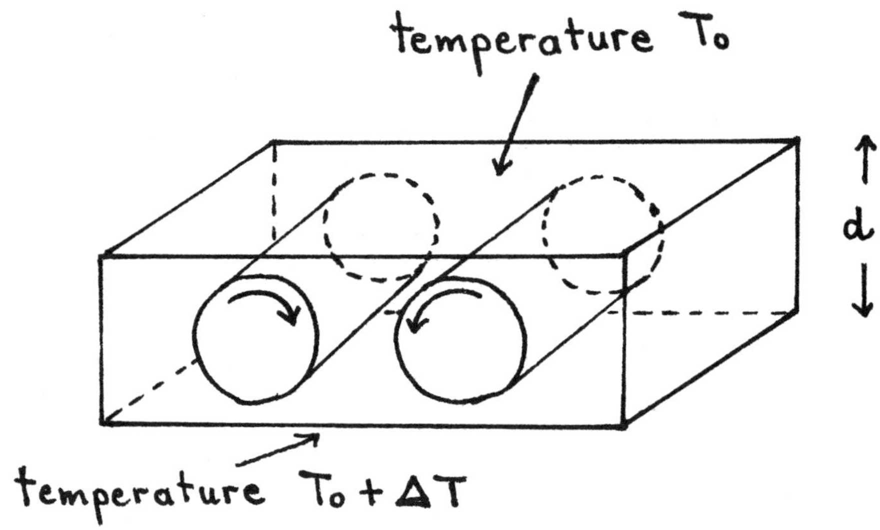

Figure 3.4 Convective motion of a fluid in the Rayleigh-Bénard configuration. Two convection cells are shown.

An ingenious theoretical analysis of the transition from normal heat conduction to convective flow was developed about a hundred years ago by Rayleigh. The theory is formulated with the help of a dimensionless number, now known as the Rayleigh number, proportional to $\Delta T d^3$. If the Rayleigh number Ra exceeds a critical value Ra_c approximately equal to 1700, the fluid is predicted to go into convective motion. Experiments on different fluids are in good agreement with the predicted value. (For more details see, for example, Tritton, 1988.)

Driven by an interest in the problem of long-range weather forecasting, the meteorologist E.N. Lorenz approached the Rayleigh-Bénard convection problem in a different way from Rayleigh's treatment. Starting from the fundamental equations of fluid dynamics, which express Newton's second law plus conservation of mass and energy, Lorenz reduced them to a form appropriate to the Rayleigh-Bénard configuration. The partial differential equations that he obtained could be considered to give rise to an infinite set of harmonics— the components of a Fourier series—just as the partial differential equation for the vibration of a string yields an infinite set of harmonics. An important difference between the vibrating string and convective flow, however, is that the string equation is linear while the convective equations are nonlinear. Lorenz simplified his equations drastically by keeping only the three most important terms in the Fourier expansion. This led him to the following relatively small set of differential equations (Lorenz, 1963):

$$\begin{aligned}\frac{dx}{dt} &= -\sigma x + \sigma y \\ \frac{dy}{dt} &= -xz + rx - y \\ \frac{dz}{dt} &= xy - bz.\end{aligned} \tag{3.6}$$

To gain an appreciation of these equations, which exhibit the transition to convective flow, it is not necessary to have a complete understanding of the variables x, y and z and the parameters r, σ and b. It is enough to say that $x(t)$ and $y(t)$ are Fourier amplitudes and that $z(t)$ describes how the vertical temperature profile varies from a linear relation. The most important parameter in the Lorenz equations is $r = Ra/Ra_C$; convection begins when r exceeds unity. The other dimensionless parameters are σ, the Prandtl number of fluid dynamics, and b which is a ratio of two lengths describing the convective cells. In computing solutions to these equations it is customary to set $\sigma = 10$ and $b = 8/3$.

What makes the Lorenz equations of great interest—far beyond the problem in fluid dynamics from which they arose—is the character of the solutions as the value of the *control parameter* r changes. This will now be described in some detail.

3.6 Equilibrium points of the Lorenz equations

The set of differential equations (3.6) is nonlinear because of the presence of the terms xz and xy. As happens for almost all nonlinear differential equations, a solution cannot be written down in a neat mathematical form. Nevertheless, given initial values $x(0)$, $y(0)$, $z(0)$, one can solve the equations as a function of time using straightforward techniques of numerical integration.

Before describing the solutions it is worth noting that if the three time derivatives are set equal to zero, the resulting equations for x, y and z have the following solutions, denoted by $\bar{x}$, $\bar{y}$ and $\bar{z}$:

$$\begin{aligned}&\text{(i)} \quad \bar{x} = 0, \quad \bar{y} = 0, \quad \bar{z} = 0; \\ &\text{(ii)} \quad \bar{x} = \bar{y} = +[b(r-1)]^{1/2}, \quad \bar{z} = r - 1; \\ &\text{(iii)} \quad \bar{x} = \bar{y} = -[b(r-1)]^{1/2}, \quad \bar{z} = r - 1.\end{aligned} \tag{3.7}$$

Each of these solutions is called an *equilibrium point* or a *fixed point* of the set of differential equations. The name is appropriate since if the system were to start out at one of these points in the xyz-space, it would remain forever at that point because the three time derivatives are zero.

Now suppose that the control parameter r has a value in the range between 0 and 1. Accurate numerical solutions show that no matter what initial values are given to x, y and z, the resulting trajectory always rapidly approaches the equilibrium point (i) of equations (3.7). Physically this means that when $0 < r < 1$ any initial disturbance represented by $x(0)$, $y(0)$, $z(0)$ dies away with time leaving the fluid at rest between the two plates. The point $\bar{x} = \bar{y} = \bar{z} = 0$ is a point attractor.

When $r = 1$ the three equilibrium points in (3.7) coincide, but once r becomes greater than unity the character of the solutions of the differential equations changes. Given some initial values for x, y and z, one finds that the numerical solution for the trajectory rapidly approaches one or other of the equilibrium points (ii) or (iii) of equations (3.7). These two points, which are point attractors, represent steady-state convection, one in the clockwise sense, the other counterclockwise. This situation prevails for r in the range between 1 and 24.74.

3.7 The Lorenz attractor

The character of the solutions changes dramatically once r exceeds 24.74. In his original paper Lorenz (1963) reported numerical results for $r = 28$. With the help of modern computer graphics the trajectories can be displayed in three dimensions as in Figure 3.5. After an initial transient the trajectories appear to approach a

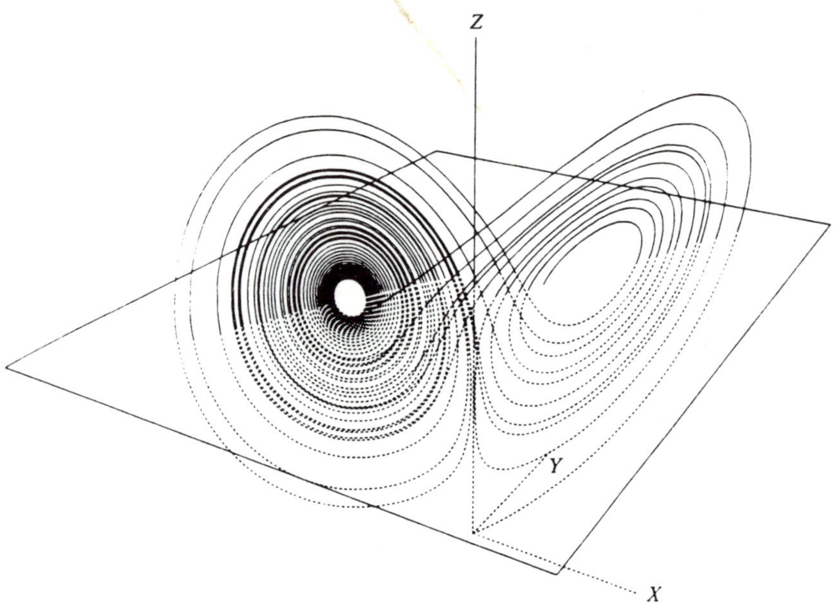

Figure 3.5 A trajectory on the Lorenz attractor for $r = 28$. From Lanford (1977).

surface consisting of two lobes. At the centre of each lobe is one of the equilibrium points (ii) and (iii) of equations (3.5). Now these points are no longer attractors. A typical trajectory spirals outward on one of the lobes until at some point it crosses over and is injected into the middle region of the other lobe. It then spirals outward on that lobe for some number of turns until it crosses back to the middle region of the other lobe. The process of spiralling outward and crossing over repeats indefinitely with varying numbers of turns each time. Although the spirals appear to lie on two-dimensional surfaces, careful numerical calculations show that these 'surfaces' consist of many sheets or leaves— in fact, an infinite number. By means of this complex geometrical structure a single trajectory succeeds in never crossing itself or closing upon itself. If a straight line is drawn approximately perpendicular to one of the lobes, it intersects the infinity of sheets in points that constitute a Cantor set similar to but not exactly the same as that described in section 2.5. The many-leaved two-headed monster which the trajectories approach asymptotically when $r > 24.74$ is the Lorenz attractor. Calculations by the box-counting algorithm described in sections 2.4 and 2.5 are not feasible for the Lorenz attractor, but its fractal dimension has been determined by other means to be 2.06 ± 0.01 (Greenside *et al.*, 1982). Because it has an infinite number of sheets but, nonetheless, occupies zero volume in the three-dimensional space, its fractal dimension is slightly greater than 2.

3.8 Determinism and unpredicability

Consider a particular trajectory running close to the surface of the Lorenz attractor. If one were to make a list of the number of turns it makes between crossovers, one would find that the numbers appear to have no particular pattern and in that sense seem to be chaotic. Nevertheless, the trajectory is completely deterministic. If two people in different countries using different high-powered computers and different numerical integration routines were to start their calculations of the Lorenz equations from exactly the same initial values $x(0)$, $y(0)$ and $z(0)$, then, provided the step-length of both numerical integrations is chosen sufficiently small to ensure high numerical accuracy, the computed trajectories, including the number of turns on alternate lobes, would be found to be the same (although not forever). Nevertheless, this apparent predictability of the trajectories is illusory for a very interesting reason.

Lorenz discovered by accident (see Gleick, 1987) that when $r = 28$, if two trajectories start out from points $x_1(0)$, $y_1(0)$, $z_1(0)$ and $x_2(0)$, $y_2(0)$, $z_2(0)$ such that their respective coordinates are very nearly the same but differ in the fifth or sixth significant figures, the two computed trajectories appear to be close together for a while but then move rapidly away from each other. Further investigations showed that initially close trajectories always diverge from each other exponentially as a function of time. If $\delta(0)$ is a tiny separation between two initial points at $t = 0$, and if $\delta(t)$ is the distance between the corresponding points on the two trajectories at the time t, calculations show that for times that are not too long,

$$\delta(t) = \delta(0)\exp(\lambda t), \tag{3.8}$$

where λ is approximately 0.9 (Bergé *et al.*, 1984, p.129). Equation (3.8) is an accurate description of the trajectories until $\delta(t)$ has grown to be a size approaching that of the Lorenz attractor itself. (Because the Lorenz attractor is bounded in xyz space, the exponential divergence cannot continue forever.) The parameter λ, which determines how rapidly the trajectories diverge, is called the ***Lyapunov exponent.*** More precisely, it is the ***largest*** Lyapunov exponent, but we need not be concerned with the other members of the family of Lyapunov exponents since they do not determine the exponential divergence of general trajectories..

The property expressed by equation (3.8) is often described as ***extreme sensitivity to initial conditions.*** Lorenz realized that this property of the solutions of his differential equations for $r = 28$ makes predictability over long times impossible because in practice there is a limit to how precisely the initial values $x(0)$,

$y(0)$, $z(0)$ can be specified. If those two distant people sitting at their different computers were to feed the same values of $x(0)$, $y(0)$ and $z(0)$ to 16 significant figures into their programs and were to use exactly the same numerical integration routines, the computed trajectories would eventually depart from each other as a result of tiny numerical differences arising from the different computer architectures. The equations are deterministic but at the same time the solutions are ultimately unpredictable for long times.

The situation is even more unpredictable when a physical system is modelled with the help of experimental measurements. If the model employs a set of nonlinear differential equations having the property of extreme sensitivity to initial conditions, prediction is possible only for relatively short times. Believing that the Earth's atmosphere is just such a system, Lorenz wrote in the concluding section of his 1963 paper: "In view of the inevitable inaccuracy and incompleteness of weather observations, precise very-long-range forecasting would seem to be nonexistent."

3.8 Chaos defined

We are now in a position to say exactly what we mean by chaotic behaviour in a dissipative dynamical system. Such a system is chaotic if the equations describing it have the property of exponentially diverging trajectories, as expressed by equation (3.8). This implies that the largest Lyapunov exponent is positive and nonzero for all trajectories.

Since Lorenz's landmark paper in 1963 scientists in many disciplines have discovered models which exhibit chaotic behaviour. A model of interest to earth scientists, which actually predates Lorenz's work, is the Bullard-Rikitake dynamo model of the Earth's magnetic field (Bullard 1955, Rikitake 1958). It turns out that the Rikitake equations have a form identical to the Lorenz equations: whenever the system trajectory crosses over from one lobe to the other, the Earth's magnetic field reverses its direction.

References

Bergé, P., Pomeau, Y., and Vidal, Ch., 1984, Order within chaos, Wiley, New York, and Hermann, Paris.

Bullard, E.C., 1955, The stability of a homopolar dynamo, Proc. Camb. Phil. Soc., v.51, p.744-760.

Gleick, J., 1987, Chaos: making a new science, Viking, New York.

Greenside, H.S., Wolf, A., Swift, J., and Pignataro, T., 1982, Impracticality of a box-counting algorithm for calculating the dimension of strange attractors, Phys. Rev. A, v.25, p.3453-3456.

Lanford, O., 1977, Turbulence Seminar, edited by P. Bernard and T. Ratiu, Lecture Notes in Mathematics No. 615, Springer, Berlin, p.114.

Lorenz, E.N., 1963, Deterministic nonperiodic flow, Journal of Atmospheric Sciences, v.20, p.130-141.

3.11

Rikitake, T., 1958, Oscillations of a system of disk dynamos, Proc. Camb. Phil. Soc., v.54, p.89-105.

Tritton, D.J., 1988, Physical fluid dynamics, 2nd ed., Oxford UP, Oxford and New York.

van der Pol, B., 1927, Forced oscillations in a circuit with nonlinear resistance (receptance with reactive triode), London, Edinburgh and Dublin Phil. Mag., v.3, p.65.

CHAPTER 4

CHAOS IN A TIME SERIES

David Goodings
Physics Department
McMaster University
Hamilton, Ontario
L8S 4M1

4.1 Introduction

In many branches of science experimental data often take the form of repeated measurements of a single variable. Examples are the set of daily maximum or minimum temperatures at a particular place; measurements of a velocity component at a particular position in a turbulent fluid, taken at intervals of, say, 6 ms; daily measurements of the concentration of carbon dioxide in the atmosphere at a particular location. When arranged chronologically into a time series, such measurements may vary in a regular fashion (as in the case of the carbon dioxide measurements at Mauna Loa on the Island of Hawaii) or may be quite irregular while displaying a slowly-varying regular component. Fourier analysis of such a time series may reveal components with well-defined periods, a well known example being the climate cycles deduced from the oxygen isotopic ratios in ocean sediments. However, Fourier techniques are not very informative if the time series is quasiperiodic (meaning nearly but not exactly periodic) or if it has the character of $1/f$ noise (as in the incidence of earthquakes exceeding a given magnitude).

A different approach to analyzing a time series based on nonlinear dynamics is the subject of this chapter. It is inspired by the possibility that, despite their irregular appearance, the measurements in a time series may be far from random or stochastic since they originate in processes that are strongly deterministic. Just as the Lorenz equations discussed in the last chapter exhibit 'deterministic chaos' when $r = 28$, it is possible that the measurements in a time series are governed by processes which can be modelled by a relatively small number of differential equations. How can one test for this possibility? Is it feasible to estimate the number of essential variables required to model the behaviour of the system?

4.2 The Santa Cruz conjecture

In 1980 a group of graduate students in physics at the University of California at Santa Cruz devised a practical method for searching for evidence of deterministic chaos in a time series (Packard *et al.*, 1980). Their approach is based on the supposition that the measurements reflect the behaviour of a dynamical system which is well described by a relatively small number of ordinary differential equations, and that associated with this set of differential equations there is an attractor whose geometrical form is unknown. The attractor may be simple, such as a limit cycle or a torus, or it may be complicated with a nonintegral fractal dimension, as in the case of the Lorenz attractor.

Packard *et al.* illustrated their method by considering the following set of ordinary differential equations:

$$\begin{aligned}\frac{dx}{dt} &= -(y+z),\\ \frac{dy}{dt} &= x+0.2y,\\ \frac{dz}{dt} &= 0.4+xz-5.7z.\end{aligned} \tag{4.1}$$

First introduced by Rössler (1976), these equations are known to show chaotic behaviour. If they are integrated numerically starting from some arbitrary initial point, the resulting trajectory projected onto the x-y plane (simply ignoring the z-component) looks like that shown in Figure 4.1. If one accepts the mathematicians' assertion that the solution of such a set of differential equations is unique at every point and therefore cannot intersect itself, one can see that Figure 4.1 is the projection onto the x-y plane of a trajectory lying on or very close to a surface in 3-dimensional space. Furthermore, this surface appears to be twisted in a manner reminiscent of a Möbius strip. In fact, the Rössler attractor is just such a twisted surface, consisting of many sheets or leaves like the Lorenz attractor.

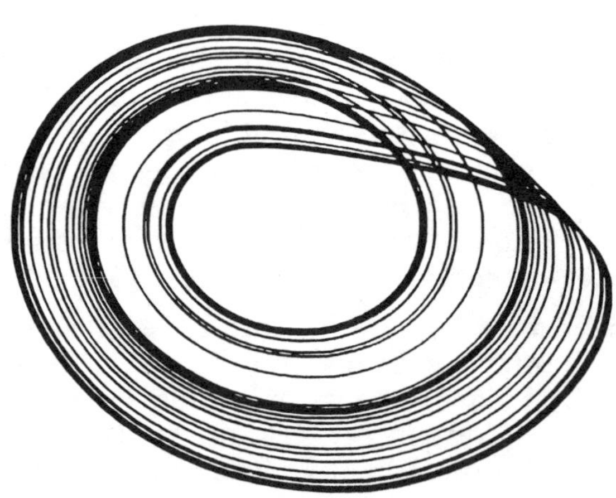

Figure 4.1 A typical trajectory of the Rössler equations projected onto the x-y plane. From Packard *et al.* (1980).

4.3

From the numerical solution of the Rössler equations one can construct a time series of one of the variables, say $x(t)$. Suppose the elements of the time series are the values $x_1, x_2, \ldots, x_N$ calculated at the times $t_i = t_0 + i\Delta t$, $i = 1, \ldots, N$. The time interval Δt between successive elements might be chosen to be about 0.1 times the time for the trajectory to make one circuit around the centre of Figure 4.1. Provided the number of entries N is reasonably large (say about 500 or more), can one somehow use the time series to 'reconstruct' the geometrical features characterizing the Rössler attractor?

At first sight it might be thought that this would be impossible since there is no explicit information in the time series about the variables $y(t)$ and $z(t)$. However, the Santa Cruz group made the bold hypothesis that to reconstruct the geometry of the attractor it is sufficient to have information about *any* three 'independent' variables, where the meaning of the word independent will be clarified later in this section. An example of such a set of independent quantities is $x(t)$ together with its derivatives dx/dt and d^2x/dt^2.

Now a numerical approximation to dx/dt at the time t_i can be constructed from the time series using any of the expressions,

$$\frac{x_{i+1} - x_i}{\Delta t}, \qquad \frac{x_i - x_{i-1}}{\Delta t}, \qquad \text{or} \qquad \frac{x_{i+1} - x_{i-1}}{2\Delta t}. \tag{4.2}$$

Likewise, a numerical approximation to d^2x/dt^2 at time t_i can be constructed from,

$$\left(\frac{x_{i+1} - x_i}{\Delta t} - \frac{x_i - x_{i-1}}{\Delta t}\right)\frac{1}{\Delta t} = \frac{x_{i+1} - 2x_i + x_{i-1}}{\Delta t^2}. \tag{4.3}$$

Packard *et al.* tested their conjecture by using a time series for $x(t)$, calculated from the Rössler equations at closely spaced time intervals, to make a plot of x and an approximation to dx/dt instead of x and y. This 'reconstructed trajectory' in the space $(x, dx/dt)$ is shown in Figure 4.2. It beautifully displays the main features of Figure 4.1 including the twist seen in projection.

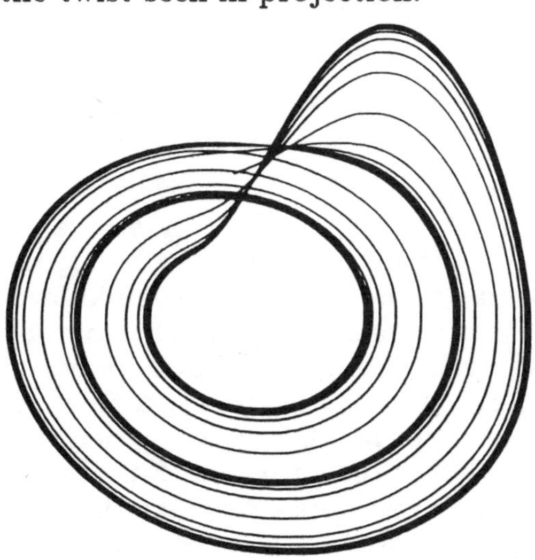

Figure 4.2 A 'reconstructed trajectory' from the time series for x(t), projected onto the x-dx/dt plane. From Packard *et al.*, (1980).

There is, however, a much more demanding quantitative test of the validity of the conjecture of Packard *et al.*: use the reconstructed trajectory in the space $(x, dx/dt)$ to estimate the positive Lyapunov exponent of the Rössler system. With the help of a technical device called a return map which we shall not attempt to describe, they calculated a value of $\lambda = 0.0677 \pm 0.0001$, in excellent agreement with the value $\lambda = 0.0677 \pm 0.0005$ obtained by analyzing the exponentially diverging trajectories of the original Rössler equations. This surprising result gives one confidence that a time series acquired from a single variable is sufficient to reconstruct the geometrical form of the Rössler attractor in 3 dimensions. Moreover, if a quantitative feature like the value of the positive Lyapunov exponent is preserved, it should also be possible to use the time series to estimate the fractal dimension D of the attractor. How to do this is described in the next section.

Let us return briefly to the question of what is meant by saying that any three independent quantities suffice to reconstruct the geometry of an attractor, assuming it exists. The meaning is similar to saying that any three linearly independent vectors can be used to construct any vector in a 3-dimensional space. It may be noted from (4.2) and (4.3) that approximations to the first and second derivatives of $x(t)$ at time t_i involve only the three values x_{i-1}, x_i and x_{i+1} in the time series. This strongly suggests that instead of choosing the three independent quantitities to be x, dx/dt and d^2x/dt^2 one can equally well choose them to be x_{i-1}, x_i and x_{i+1}. This final step underlies the developments that follow.

4.3 The dripping tap

The method described in the last section is not restricted to mathematical models such as the Rössler equations and the Lorenz equations. An instructive and appealing example of its applicability to real physical systems has been described by the Santa Cruz group in an article in Scientific American (Crutchfield *et al.*, 1986). If the flow of water through an ordinary tap is suitably adjusted, the tap may be made to drip in a perfectly periodic fashion, or, at somewhat higher flow rates, in a manner such that the time intervals between successive drips appear to vary chaotically. In the latter flow régime, the Santa Cruz group let the drops fall onto a microphone connected to a recording device, thereby obtaining a record of the resulting sound pulses. The measured time intervals between successive peaks of the recorded sound intensity were made into the elements of a time series.

Are the time intervals between drops randomly distributed or are they governed by some sort of deterministic process? To answer this question Crutchfield *et al.* assumed the latter to be the case and used the time series to attempt to construct the geometry of a conjectured attractor of unknown form. The first three elements of the time series can be used to make the components (x_1, x_2, x_3) of a vector from the origin to the point (x_1, x_2, x_3) in 3-dimensional space. Similarly, (x_2, x_3, x_4) constitute a second vector, (x_3, x_4, x_5) a third and so on through the entire time series, yielding a total of $N-2$ vectors. When the corresponding points (x_i, x_{i+1}, x_{i+2}), $i = 1, \ldots, N-2$ were plotted on a 3-dimensional plot, they were found to lie near a twisted surface that looks very like the Rössler attractor. In contrast, if the elements of the time series had been randomly distributed (about some positive mean value), the corresponding 3-dimensional vectors would have entirely filled some volume of the space with randomly generated points. The fact that the vectors constructed from the dripping tap's time series appear to lie on a surface is an indication that the flow from the tap constitutes a nonlinear dynamical system governed by a low-dimensional chaotic attractor.

4.4 The correlation integral

We have seen how a time series of repeated measurements of a single dynamical variable can be used to reconstruct the geometry of an attractor associated with the system. We have also seen that a well-defined feature of such an attractor is its fractal dimension. It would be very satisfying if a way could be found to analyze a time series to yield an estimate of this fractal dimension. A method that under propitious circumstances will do precisely this was devised by Grassberger and Procaccia (1983). We shall refer to this as the GP method.

The object of central interest in their method is the correlation integral given by the relation,

$$C_m(r) = \frac{2}{N(N-1)} \sum_{i=1}^{N-1} \sum_{j=i+1}^{N} \theta(r - |\mathbf{X}_i - \mathbf{X}_j|), \qquad (4.4)$$

where $\theta(u)$ is the Heaviside function defined by,

$$\begin{aligned} \theta(u) &= 1 \qquad \text{for } u > 0 \\ &= 0 \qquad \text{for } u < 0. \end{aligned} \qquad (4.5)$$

In (4.4), $\mathbf{X}_i = (x_i, x_{i+1}, \ldots, x_{i+m-1})$ is an m-dimensional vector drawn from the time series beginning with the i-th element. (In the example of the dripping tap the vectors were chosen to have 3 components, but there is no need to restrict the number of components in this way.) The time series is assumed to have at least $N + m - 1$ elements, and the summations over i and j yield all pairs of distinct m-dimensional vectors, $N(N-1)/2$ in number. The distance between $\mathbf{X}_i$ and $\mathbf{X}_j$ in the m-dimensional space is calculated in the usual way:

$$|\mathbf{X}_i - \mathbf{X}_j| = \left(\sum_{k=0}^{m-1} (x_{i+k} - x_{j+k})^2 \right)^{\frac{1}{2}}. \qquad (4.6)$$

It is not hard to see from (4.4) and (4.5) that if r is greater than the distance between *any* two of the vectors, then every term in the double sum of (4.4) will be unity and hence $C_m(r)$ will be equal to 1. It is also easy to see that, provided no two vectors $\mathbf{X}_i$ are identical, $C_m(r) \to 0$ as $r \to 0$. Thus, $C_m(r)$ starts out at zero for r small and increases monotonically toward unity as r increases.

Let us assume that the elements of the time series x_1, $x_2, \ldots,$ x_{N+m-1} come from a system the dynamical behaviour of which is governed by a small number of nonlinear differential equations, and that these equations possess an attractor of unknown geometrical form. Furthermore, we shall assume henceforth that any initial transient decayed to zero before the beginning of the time series. The vectors $\mathbf{X}_i$, $i = 1, \ldots, N$ will, therefore, lie close to the surface (or hypersurface, if $m > 3$)

of a geometrical form representing the 'reconstructed attractor'. If this object exists, it is plausible that at small values of r,

$$C(r) = \text{const}\ r^{\nu}, \qquad r \text{ small}, \tag{4.7}$$

where the exponent ν is called the correlation dimension. The geometrical basis for this hypothesis is revealed by the following simple example.

Suppose the vectors $\mathbf{X}_i$ have 3 components and that the reconstructed attractor happens to be a sphere of radius R with its centre at the origin of the 3-dimensional space. Since the initial transient has been removed, each of the N vectors $\mathbf{X}_i$ joins the origin to a point on the surface of the sphere. Let us further assume that these vectors are uniformly distributed over the sphere so that the number per unit surface area is approximately constant. If we then choose a particular vector, say $\mathbf{X}_k$ corresponding to the single term $i = k$ in (4.4), and if we choose r to be small compared to the sphere radius R, then the nonzero terms contributing to $C_3(r)$ come from vectors $\mathbf{X}_j$ lying within a 'ball' of radius r having its centre at $\mathbf{X}_k$, as shown in Figure 4.3.

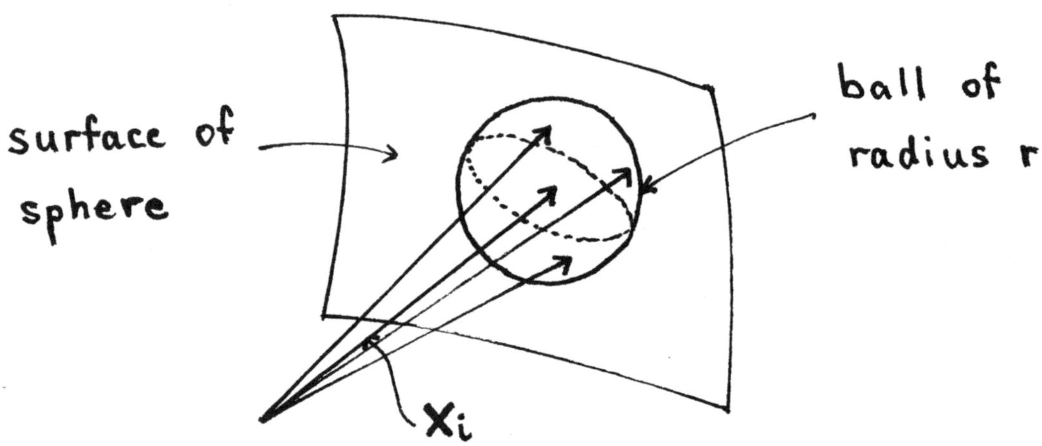

Figure 4.3 A ball of radius r centered on the vector $\mathbf{X}_i$ situated on or near the surface of a hypothesized spherical attractor. Other vectors $\mathbf{X}_j$ lying within the ball are also shown.

How many vectors $\mathbf{X}_j$ will there be such that $|\mathbf{X}_k - \mathbf{X}_j| < r$? Since the vectors are assumed to be uniformly distributed over the surface of the sphere, the number within the ball of radius r will be proportional to the area πr^2 of the circle of intersection of the ball with the larger sphere. The same would be true, of course, for any of the vectors $\mathbf{X}_i$ in equation (4.4). Thus, in the special case of a (hypothesized) spherical attractor, the number of pairs of vectors $\mathbf{X}_i$ and $\mathbf{X}_j$ contributing to $C_3(r)$ is proportional to r^2, when r is small compared to R. The correlation dimension ν would be 2 in this case, reflecting the two-dimensional character of the sphere's surface. Clearly this argument would not be altered if the reconstructed attractor were an ellipsoid or any other surface obtained by deforming a sphere in a reasonable manner. (What matters is that the ball of radius r, centered on any vector, cuts out such a small section of the attractor's surface that it is close to being flat.)

4.7

Thus, if the reconstructed attractor is a well-behaved two-dimensional surface and if the vectors $\mathbf{X}_i$ from the time series are distributed uniformly over it, the correlation integral should be proportional to r^2 for small values of r. This conclusion remains true when the number of components of the vectors is greater than 3.

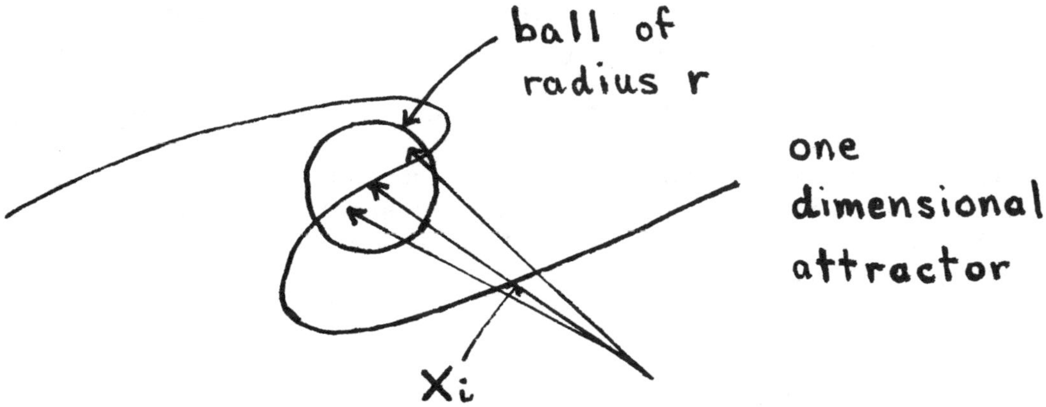

Figure 4.4 A ball of radius r centered on the vector $\mathbf{X}_i$ situated on or near a hypothesized one-dimensional attractor. Other vectors $\mathbf{X}_j$ lying within the ball are also shown. After Bergé *et al.* (1984).

In a different example, suppose the reconstructed attractor is a meandering one-dimensional curve, as shown in Figure 4.4, with the vectors $\mathbf{X}_i$ uniformly distributed along it. An argument similar to that of the preceding paragraphs would lead to the conclusion that $C_m(r)$ is proportional to r for small r. The correlation dimension ν would be 1 in this case, reflecting the one-dimensional form of the attractor. Hence if the attractor is not strange, its correlation dimension is just what one would expect it to be.

What happens if the reconstructed attractor *is* strange (meaning that it has a fractal dimension D), as in the case of the Lorenz attractor? Even if the vectors of the time series were uniformly distributed over the 'surface', it is not obvious how the calculated correlation dimension ν would be related to the fractal dimension D of the attractor. However, Grassberger and Procaccia (1983) were able to show that, no matter what its geometrical form is, an attractor always has $D \geq \nu$. In other words, the correlation dimension provides a lower bound for the fractal dimension of the reconstructed attractor.

Grassberger and Procaccia calculated the correlation dimension for several time series constructed from mathematical models. In the case of a time series of 15,000 elements obtained from a numerical solution of the Lorenz equations (equations (3.4)), with $\Delta t = 0.25$, they found $\nu = 2.05 \pm 0.01$, in satisfactory agreement with the best estimate for the dimension of the Lorenz attractor, $D = 2.06 \pm 0.01$. In three other examples, chosen because they have low-dimensional attractors, the calculated values of ν were found to be close to but less than the known values of D, as expected. These results indicate that the GP method may be useful in estimating a lower bound for the fractal dimension of a possible attractor associated with an experimental time series.

4.5 Evidence for an attractor in Rayleigh-Bénard convection

One of the first successful applications of the GP method to a time series of experimental measurements was carried out by Malraison, Atten, Bergé and Dubois (Malraison *et al.*, 1983). They studied Rayleigh-Bénard convection in an unspecified 'test fluid' having a Prandtl number of 40. The fluid, confined in a rectangular box, showed two convective rolls in one horizontal direction and either one or two rolls in the perpendicular horizontal direction. A time series was obtained from measurements of the deflection of a narrow light beam crossing the cell in the narrower horizontal direction. On increasing the temperature difference ΔT between the upper and lower faces of the box (equivalent to increasing the Rayleigh number Ra), Malraison *et al.* found a régime of periodic behaviour followed by a period-doubling cascade (to be described in Chapter 5). At still higher ΔT they found chaotic behaviour, and it was in this régime, with $Ra/Ra_c = 235$, that they collected the 15,000 elements comprising their time series.

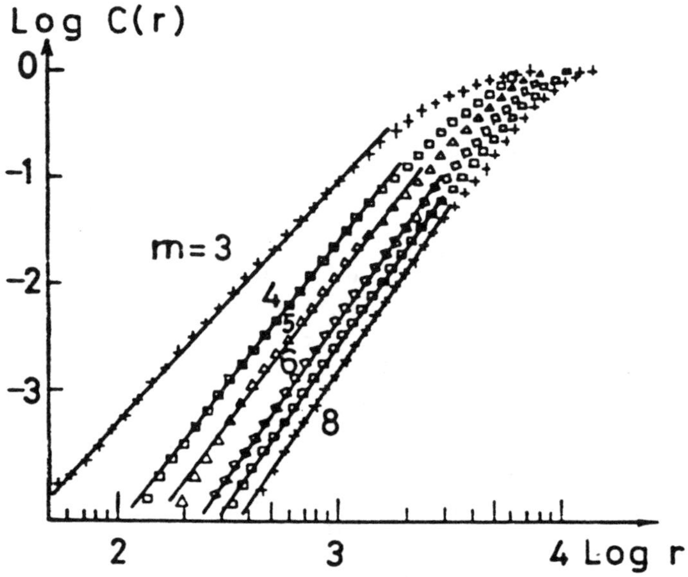

Figure 4.5 Plots of $\log C_m(r)$ against $\log r$ calculated by Malraison *et al.* (1983) for different values of the embedding dimension m.

Malraison *et al.* calculated the correlation integral of Grassberger and Procaccia on the basis of 100 vectors $\mathbf{X}_i$, rather than the $N-1$ vectors indicated in equation (4.4). Their results for $\log C_m(r)$ *versus* $\log r$ are shown in Figure 4.5. The different curves correspond to the embedding dimension m going from 2 to 11. The slopes obtained from the lower ends of these curves yield the exponent ν as a function of m. This is shown in Figure 4.6. It may be seen that ν rises with m up to $m = 4$ but thereafter remains approximately constant at a value of 2.8. This points to the possible existence of a strange attractor associated with the chaotic convective flow in their system having a fractal dimension $D \approx 2.8 \pm 0.1$.

As a check on their work Malraison *et al.* performed similar calculations for a contrasting case, namely a time series of 15,000 random numbers generated so as to span the same range of numerical values as their experimental data. For given

4.9

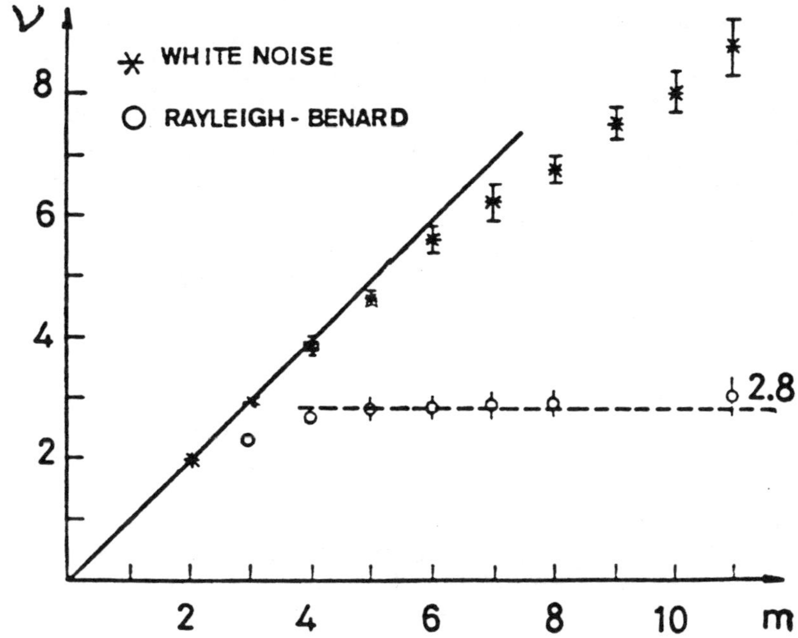

Figure 4.6 The exponent ν deduced from the curves of Figure 4.5 plotted against the embedding dimension m. The points shown as asterisks were calculated from a time series of random numbers. After Malraison *et al.* (1983).

embedding dimension m, the m-dimensional vectors constructed from the time series are expected to fill the entire space. Consequently, the exponent ν calculated from the curve of $\log C_m(r)$ *versus* $\log r$ would be expected to be close to m. This expectation is fufilled with good accuracy when $m \leq 4$, as may be seen by the points shown as asterisks in Figure 4.6. For $m > 4$, however, the calculated values of ν gradually fall away from the straight line $\nu = m$, although they continue to rise steadily as m increases. This behaviour undoubtedly reflects the limited size of the data set and the steadily deteriorating statistics of the distribution of vectors $\mathbf{X}_i$ as m increases.

Calculations based on a time series of random numbers have become a standard accessory to the GP analysis of experimental time series. They show very clearly the dimension at which the statistics become degraded to the point that one loses confidence in the subsequent calculated values of ν.

Another check carried out by Malraison *et al.* should be mentioned. Successive elements in the time series $x_1, x_2, \ldots, x_N$ have been assumed to be separated by the time interval Δt. However, there is no compelling reason to use consecutive elements of the time series in constructing m-component vectors, as we have been assuming until now. Instead, one can choose a *time delay* T between successive components of a vector to be some multiple of Δt, for example, $T = p\Delta t$, where p is a positive integer. Malraison *et al.* carried out calculations of the correlation integrals $C_m(r)$ with p varying from 2 up to 15. They concluded that the calculated exponent ν is, to a first approximation, independent of the time delay for $2 < p \leq 10$.

4.6 Critique of the GP method

The method of Grassberger and Procaccia for analyzing a time series of experimental measurements scored some notable successes in the years immediately following its proposal. In addition to the work of Malraison *et al.* (1983) described in the last section, the GP method was successfully applied by Brandstäter *et al.* (1983) to a velocity time series measured in Couette flow. They showed that at the onset of turbulence the correlation dimension begins to rise above a value near 2, suggesting that a low-dimensional chaotic attractor exists in the turbulent flow régime.

Despite these successes, the GP method has been found to have a number of problems which have generated an extensive literature. The purpose of this section is to acquaint the reader with some of these difficulties and to indicate how some, at least, may be overcome.

First of all, the correlation dimension ν may be found to approach a relatively small nonintegral value as the embedding dimension m increases, *even when the time series has been generated by means of nondeterministic random signals* (Osborne and Provenzale, 1989). Thus, it is possible to be misled into thinking that a system is governed by some underlying deterministic process when, in fact, it is responding to random inputs. This is a major problem which implies that, by itself, the correlation dimension is insufficient to permit conclusions to be drawn about the existence or nonexistence of a strange attractor.

An important consideration in analyzing a time series of experimental measurements is the effect of experimental uncertainties or 'noise'. This has been investigated by Ben Mizrachi, Procaccia and Grassberger (1984) who showed that when the correlation integral $C_m(r)$ is computed for a noisy time series and $\log C_m(r)$ is plotted against $\log r$, the slope of the curves at the lowest values of $\log r$ is very close to the embedding dimension (for $m \leq 5$). This corresponds to random noise being 'space filling'. However, a change in slope of the curves of $\log C_m(r)$ *versus* $\log r$ occurs when r exceeds the size of the random uncertainties in the data. The interval of $\log r$ above this 'break point', showing a straight-line behaviour, is called the *scaling region*, and the slope over this interval is taken to be the correlation dimension ν. Clearly it is very important to have some way of estimating the experimental uncertainties in a time series in order to know approximately where the scaling region begins.

In practice, the scaling region may turn out to be small or nonexistent. (See, for example, Cannizzo and Goodings, 1988; Lochner *et al.,* 1989.) Unfortunately, increasing the embedding dimension m in order to check the convergence of ν has the effect of causing the scaling region to shrink. A method devised to minimize the required embedding dimension has been devised by Broomhead and King (1986). Based on a 'singular value decomposition', it leads to a statistically independent set of variables spanning the embedding space. A more recent paper employing this useful technique has been published by Albano *et al.* (1988).

Related to this question of independence of elements in the time series is a paper by Theiler (1986) in which it is shown that certain spurious correlations (which give a misleading value for ν) may be avoided in computing the correlation integral. The trick is to start the sum over j in equation (4.4) not at $i+1$ but at $i+k$ where k is a positive integer large enough to exclude pairs of vectors that are within a 'characteristic time' (roughly determined by the first zero in the autocorrelation function). In fact, if the time delay T, introduced at the end of the last section, is chosen to be roughly equal to the time at which the autocorrelation function first becomes zero, as suggested by Simm *et al.* (1887), then the components of each vector should be 'independent' in the sense used above, and Theiler's spurious correlations should not occur even with $k = 1$.

Another question of great interest is: How large a data set is required to provide convincing evidence for the existence of an attractor—chaotic or nonchaotic—having a given dimension D? In a general way it is clear that the larger the value of D, the more m-dimensional vectors $\mathbf{X}_i$ (where $m \approx D$) are required to pin down the geometry of the attractor in the m-dimensional space. There are many papers in the literature which report calculations of $C_m(r)$ to absurdly high values of m. In the author's opinion, if the correlation dimension ν, when plotted against m, does not quickly approach a constant value less than about 5, the GP method is likely to be quite unreliable, even when the time series consists of tens of thousands of elements. The problem of how large a data set is required to determine a correlation dimension of a given size has recently been investigated by Nerenberg and Essex (1990) who introduce the concept of a *critical embedding dimension* depending on ν.

Finally, it should be mentioned that a somewhat different method of calculating the dimension of an attractor has been proposed by Termonia and Alexandrowicz (1983), but this failed to give reasonable results when tested on a time series calculated from the Lorenz equations (Cannizzo 1989, private communication). The importance of estimating the uncertainty in any kind of dimension calculation has been stressed by Holfzfuss and Mayer-Kress (1986). They suggest methods of error estimation appropriate to the GP method, the method of Termonia and Alexandrowicz and their own 'pointwise method'.

References

Albano, A.M., Muench, J., Schwartz, C., Mees, A.I., and Rapp, P.E., 1988, Singular-value decomposition and the Grassberger-Procaccia algorithm: Physical Review A, v.38, p.3017-3026.

Ben Mizrachi, A., Proccacia, I., and Grassberger, P., 1984, Characterization of experimental (noisy) strange attractors: Physical Review A, v.29, p.975-977.

Bergé, P., Pomeau, Y., and Vidal, Ch., 1984, Order within chaos, Wiley, New York, and Hermann, Paris.

Brandstäter, A., Swift, J., Swinney, H.L., Wolf, A., Farmer, J.D., Jen E., and Crutchfield, P.J., 1983, Low-dimensional chaos in a hydrodynamic system: Physical Review Letters, v.51, p.1442-1445.

Broomhead, D.S., and King, G.P., 1986, Physica D, v.20, p.217.

Cannizzo, J.K., and Goodings, D.A., 1988, Chaos in SS Cygni? The Astrophysical Journal, v.334, p.L31-L34.

Crutchfield, J.P., Farmer, J.D., Packard, N.H., and Shaw, R.S., 1986, Chaos: Scientific American, December 1986, p.46-57.

4.12

Holzfuss, J., and Mayer-Kress, G., 1986, An approach to error estimation in the application of dimension algorithms, *in* Dimensions and entropies in chaotic systems, G. Mayer-Kress, ed., Springer-Verlag, Berlin.

Lochner, J.C., Swank, J.H., and Szymkowiak, A.E., 1989, A search for dynamical attractors in Cygnus X-1: The Astrophysical Journal, v.337, p.823-831.

Malraison, B., Atten, P., Bergé, P., and Dubois, M., 1983, Dimension of strange attractors: an experimental determination for the chaotic régime of two convective systems: Journal de Physique Lettres, v.44, p.L897-L902.

Nerenberg, M.A.H., and Essex, C., 1990, Correlation dimension and systematic geometric effects: Physical Review A, v.42, p.7065-7074.

Osborne, A.R., and Provenzale, A., 1989, Finite correlation dimension for stochastic systems with power-law spectra: Physica D, v.35, p.357-381.

Packard, N.H., Crutchfield, J.P., Farmer, J.D., and Shaw, R.S., Geometry from a time series: Physical Review Letters, v.45, p.712-716.

Rössler, O.E., 1976, An equation for continuous chaos: Physics Letters, v.57A, p.397-398.

Simm, C.W., Sawley, M.L., Skiff, F., andPochelon, A., 1987, Helvetica Physica Acta, v.60, p.510.

Termonia, T., and Alexandrowicz, Z., 1983, Fractal dimension of strange attractors from radius versus size of arbitrary clusters: Physical Review Letters, v.51, p.1265-1268.

Theiler, J., 1986, Spurious dimension from correlation algorithms applied to limited time-series data: Physical Review A, v.34, p.2427-2432.

CHAPTER 5

ROUTES TO CHAOS

David Goodings
Physics Department
McMaster University
Hamilton, Ontario
L8S 4M1

5.1 Introduction

Until now we have not considered in detail how irregular or chaotic motion evolves out of regular motion as a control parameter of a system is varied. In describing the experiments of Malraison *et al.* (1983) in section 4.5, it was mentioned that as one increases the temperature difference ΔT between the upper and lower plates in the Rayleigh-Bénard configuration, one finds first a régime of periodic behaviour followed by a period-doubling cascade. When ΔT is further increased, the fluid undergoes a transition to turbulence—the classic example of chaotic behaviour in a physical system. In this chapter we describe a mathematical model investigated by M.J. Feigenbaum which shows this period-doubling behaviour and the transition to chaos. This is one of the possible routes to chaos. Other possibilities are briefly described at the end of the chapter.

5.2 Period doubling in Rayleigh-Bénard convection

Libchaber and Maurer (1980) have carried out a series of unusual experiments on liquid helium in the Rayleigh-Bénard configuration (see Section 3.5). Their apparatus was tiny with horizontal plates measuring 3.0 mm by 1.25 mm and a separation between the hotter and colder plates of 1.5 mm. Two convective cells formed in the box. Instruments were positioned in the upper plate above each of the convective cells to monitor the temperature $T(t)$ as a function of time. The signal

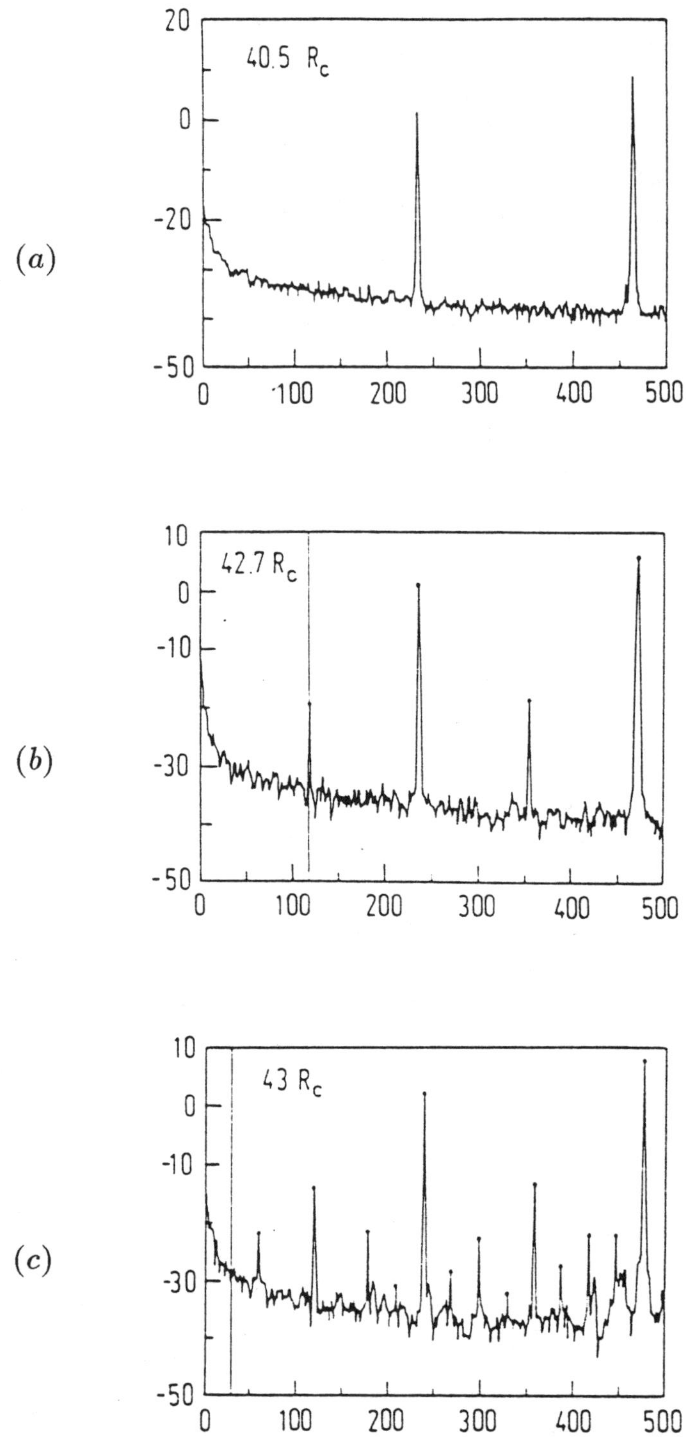

Figure 5.1 Power spectrum of the temperature $T(t)$ measured at the top face of a Rayleigh-Bénard cell by Libchaber and Maurer (1980). In (a), (b) and (c) the Rayleigh number is indicated in terms of the critical Rayleigh number R_c. Reproduced from Schuster (1984), p.63.

was then Fourier transformed to yield the power spectrum $P(\omega)$:

$$P(\omega) = \int_0^\infty |T(t)|^2 \exp(i\omega t)dt. \qquad (5.1)$$

The results are shown in Figure 5.1 for three different values of the Rayleigh number Ra, expressed in terms of the critical Rayleigh number Ra_c (see Section 3.5). Not shown in Figure 5.1 are the results for Ra below $40Ra_c$ when there is only one prominent frequency in the power spectrum at approximately 470 mHz. This is the fundamental frequency f_0. It corresponds to a fundamental period T_0 of about 2 seconds. When $Ra = 40.5Ra_c$, as in Figure 5.1(a), a second prominent peak appears in the power spectrum at approximately 235 mHz—half the fundamental frequency or double the period T_0. At $Ra = 42.7Ra_c$ new peaks appear near $f_0/4$ and $3f_0/4$ as shown in Figure 5.1(b). The period of the system's motion has now grown to $4T_0$. Finally, Figure 5.1(c) shows that when $Ra = 43Ra_c$, new peaks at $f_0/8$, $3f_0/8$, $5f_0/8$ and $7f_0/8$ appear in the power spectrum, plus some weaker peaks at odd multiples of $f_0/16$. For Ra greater than $43Ra_c$ these lower intensity peaks become submerged in a background of noise. At this point the motion is no longer strictly periodic. These beautiful experiments of Libchaber and Maurer illustrate what is called the *period-doubling route to chaos.*

5.3 The logistic equation

A model that exhibits period doublings is a mapping first introduced in 1845 by P.F. Verhulst to model the growth and decay of the population of a biological species. Consider a single species of insect, for example, confined to a closed area such as an island. Let x_j be the ratio of the maximum number of insects in year j to the maximum number ever recorded. If x_j happens to be small initially, the population may be expected to grow under normal conditions so that in year $j+1$ it has become,

$$x_{j+1} = rx_j. \qquad (5.1)$$

Here r is a constant greater than unity which represents the growth factor. This process of unimpeded growth might proceed for several years in a row, but it clearly cannot continue forever because of competition for food and other effects of overcrowding. The latter effects can be roughly taken into account by changing the mapping to,

$$x_{j+1} = rx_j - rx_j^2, \qquad (5.2)$$

where the term invoving x_j^2 is small when the population ratio is small but becomes increasingly important as x_j increases toward its maximum value of 1. The relation,

$$x_{j+1} = rx_j(1 - x_j), \qquad 0 \le x_j \le 1 \qquad (5.3)$$

is a quadratic map known as the *logistic equation.* (The term 'quadratic map' will be explained in section 5.7.)

In what follows the constant r will be regarded as a control parameter. We shall be interested in the behaviour of the map for r in the range $0 < r < 4$. (When $r > 4$ the iterates x_j can alternate in sign and $|x_j|$ can increase toward infinity.)

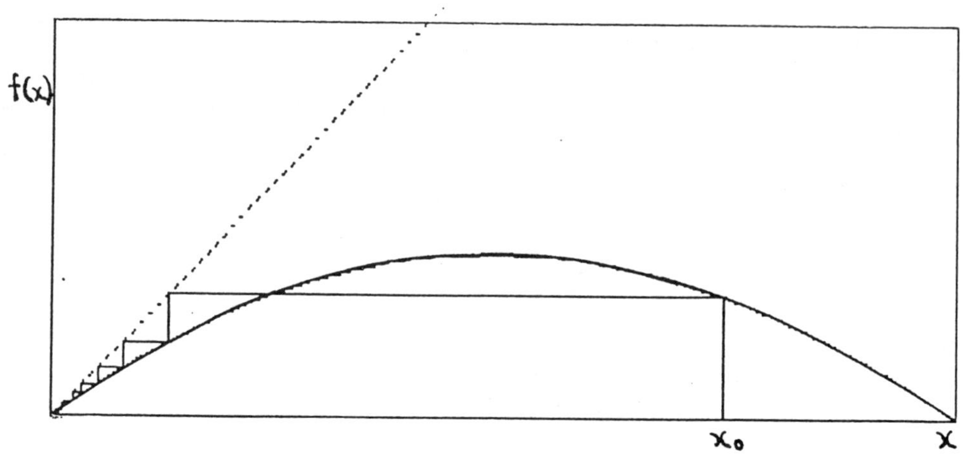

Figure 5.2 The mapping function $f(x) = rx(1-x)$ for $r = 0.7$. The dashed line is the line $y = x$ used to locate the next iterate, as explained in the text. Beginning with the initial point x_0, the iterates rapidly approach the origin.

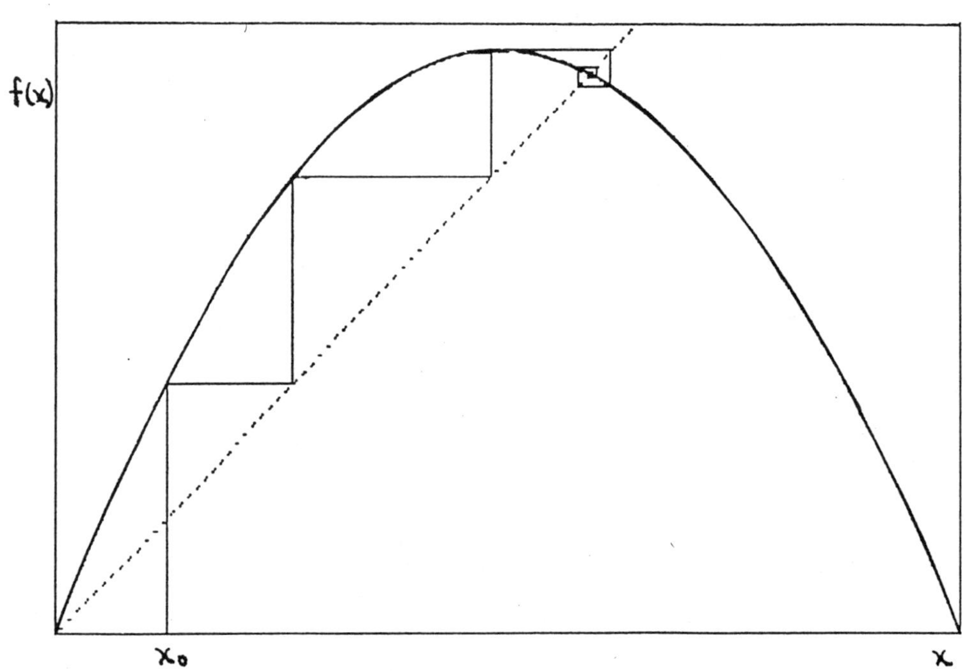

Figure 5.3 The mapping function $f(x) = rx(1-x)$ for $r = 2.5$. Beginning with the initial point x_0, the iterates approach the fixed point $x^* = 1 - 1/r = 0.60$.

5.5

To study the interesting properties of the logistic map it is helpful to define the continuous function,

$$f(x) = rx(1-x), \qquad 0 \le x \le 1. \tag{5.4}$$

This is shown in Figure 5.2. It may be seen to have a maximum value of $r/4$ at $x = 1/2$. Figure 5.2 also shows an arbitrarily chosen value x_0 and the values x_1, $x_2, \ldots$, obtained by repeatedly applying the map. The sequence of iterates is readily constructed graphically with the help of the straight line $y = x$ shown as a dashed line in Figure 5.2. For example, starting from x_0 one obtains the value of the first iterate x_1 from the height of the vertical line at x_0 shown in the figure. To locate the position x_1 along the x-axis, draw a horizontal line at height x_1 until it intersects the dashed line $y = x$, and then drop a vertical line from the point of intersection down to the x-axis. The position of x_2 can be obtained from x_1 in a similar manner. The sequence of iterates may be seen to approach the origin. Thus $x = 0$ acts like a point attractor in this case. It is also what is called a *fixed point* of the map, meaning that it maps into itself. In the present context one usually says that $x = 0$ is an *attracting fixed point* or a *stable fixed point* of the map.

When r becomes greater than 1 the mapping function $f(x)$ appears as shown in Figure 5.3, intersecting the line $y = x$ at a point away from the origin. This point, which clearly maps into itself, is a new fixed point of the map. In general the fixed points of the logistic map may be found by solving,

$$x = rx(1-x), \tag{5.5}$$

which has the solution $x = 0$ and,

$$x^* = 1 - 1/r. \tag{5.6}$$

The latter solution is the position of the point of intersection of the mapping function $f(x)$ with the straight line $y = x$. The iterates of the map, starting from any x_0 between 0 and 1, now head toward x^*, as shown by the example in Figure 5.3. It turns out that x^* is an attracting fixed point when $1 < r < 3$. In contrast, the fixed point $x = 0$ has become *unstable* since any initial point x_0 close to but not equal to zero will generate a series of iterates that move away from it, eventually approaching x^*.

5.4 The logistic equation for $3.00 < r < 3.45$

Interesting things begin to happen once the control parameter r becomes greater than 3. It turns out that when $r > 3$, both $x = 0$ and $x = x^*$ are unstable fixed points. This means that, except for the special cases when x_0 happens to be exactly equal to 0 or x^*, the iterates of the map (5.3) move away from both 0 and x^*. Where do they go?

One can answer this question by studying the map of the second iterates, i.e. taking two iterations at a time:

$$x_{j+2} = f(x_{j+1}) = f(f(x_j)). \tag{5.7}$$

This map, which goes directly from x_j to x_{j+2}, has a mapping function (analogous to equation (5.4) for the first iterates) usually denoted as $f^2(x)$ given by,

5.6

$$f^2(x) = rf(x)\{1 - f(x)\} = r[rx(1-x)]\{1 - [rx(1-x)]\}$$
$$= r^2x(1-x) - r^3x^2(1-x)^2. \qquad (5.8)$$

This is shown in Figure 5.4 for the case $r = 3.4$.

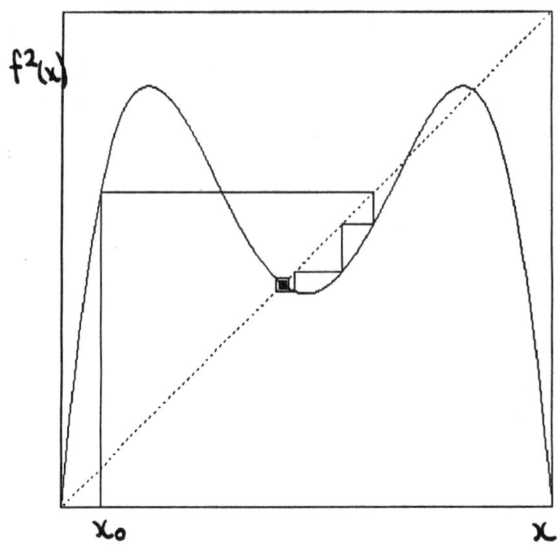

Figure 5.4 The mapping function $f^2(x)$ of the second iterates of the logistic equation for $r = 3.4$. Starting from an arbitrary point x_0, the iterates approach one of the points $\bar{x}_1$ or $\bar{x}_2$ given in equation (5.10).

The fixed points of this map are determined by the quartic equation,

$$x = r^2x(1-x) - r^3x^2(1-x)^2. \qquad (5.9)$$

Recognizing that $x = 0$ and $x = x^* = 1 - 1/r$ must be roots of this equation (since they are clearly fixed points of the map (5.7)), one can readily reduce (5.9) to a quadratic equation. With a little algebra the roots of this equation may be shown to be,

$$\bar{x}_1 = \frac{r + 1 - [(r+1)(r-3)]^{1/2}}{2r}, \qquad \bar{x}_2 = \frac{r + 1 + [(r+1)(r-3)]^{1/2}}{2r}. \qquad (5.10)$$

Now it turns out that both of these points are *attractive fixed points* of the map of the second iterates (equation 5.7). Furthermore, if one has an appetite for algebra one can show by direct substitution in equation (5.3) that,

$$\bar{x}_2 = f(\bar{x}_1), \qquad \bar{x}_1 = f(\bar{x}_2). \qquad (5.11)$$

If we go back to the original map of equation (5.3), we can construct a typical set of iterates as shown in Figure 5.5.

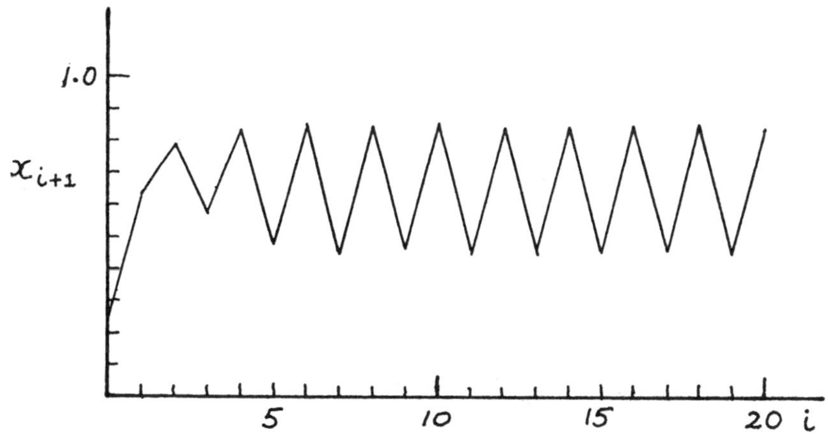

Figure 5.5 The first twenty iterates of the logistic map for $r = 3.4$.

The values jump up and down but rapidly approach the points $\bar{x}_1$ and $\bar{x}_2$. For large values of j the iterates oscillate essentially between the values $\bar{x}_1$ and $\bar{x}_2$. If one lets T_0 be the time between successive iterations of the map, after an 'initial transient' the map settles down to a periodic motion having period $2T_0$. This is the period-doubling feature of the logistic map when r changes from a value less than 3 to a value greater than 3.

5.5 The logistic equation for $3.45 < r < 3.57$

The two fixed points $\bar{x}_1$ and $\bar{x}_2$ of the map $x \longmapsto f^2(x)$ described in the last section remain attractive fixed points until r reaches the value $1 + \sqrt{6} = 3.4495$. What happens when r goes just beyond this value?

As might be expected the answer to this question is provided by considering the map of the fourth iterates,

$$x \longmapsto f^4(x) \equiv f^2(f^2(x)). \tag{5.12}$$

It is unnecessary to submit this mapping to a detailed algebraic study as computer experiments show clearly what occurs: both $\bar{x}_1$ and $\bar{x}_2$ become unstable fixed points of this map (along with $x = 0$ and $x = x^*$), while at the same time four new attractive fixed points appear. If the new fixed points are denoted $\tilde{x}_1$, $\tilde{x}_2$, $\tilde{x}_3$ and $\tilde{x}_4$, one finds that if they are properly labelled,

$$\tilde{x}_2 = f(\tilde{x}_1), \qquad \tilde{x}_3 = f(\tilde{x}_2), \qquad \tilde{x}_4 = f(\tilde{x}_3), \qquad \tilde{x}_1 = f(\tilde{x}_4). \tag{5.13}$$

The iterates of the original map, equation (5.3), rapidly approach these values, moving forever through a cycle of period $4T_0$.

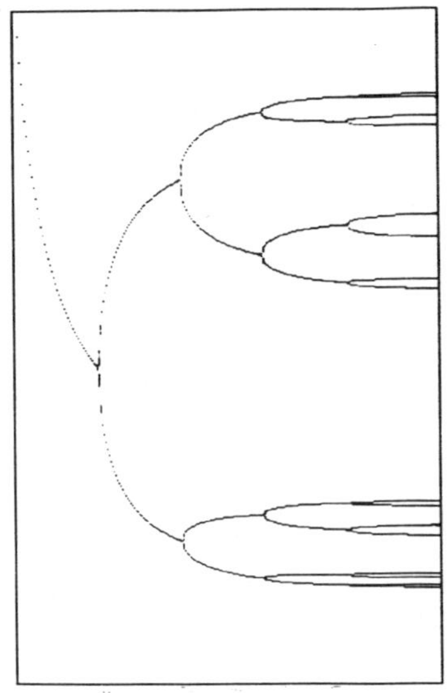

Figure 5.6 'Bifurcation cascade' for the logistic map. The figure shows the locations (x-values) of the attractive fixed points of the map as a function of $\log_{10} r$ for $3.44 < r < 3.5699$.

The pattern of period doublings continues to occur as r is further increased. Figure 5.6 shows the positions of the stable or attractive fixed points $\tilde{x}_i$ of the map $x \longmapsto f^{2^n}(x)$ as a function of r. (Note that each time a branching occurs, n must be increased by unity.) At each branching in this diagram the original fixed point becomes unstable and gives rise to two new stable fixed points. (We emphasize that the positions of only the ***stable*** fixed points are shown in Figure 5.6.) The values of r at which these ***bifurcations*** occur are called ***bifurcation values***. The first bifurcation occurs at $r_1 = 3$, the second at $r_2 = 3.45$ and so on. If the nth bifurcation occurs at r_n, it is clear from Figure 5.6 that $r_{n+1} - r_n$ decreases very rapidly as n increases. By the time r reaches the value $r_\infty = 3.5699456\ldots$, an infinite number of bifurcations have occurred and at this very special value of r there are an infinite number of attractive fixed points in the interval $0 < x < 1$.

It remains to consider the question of sensitivity to initial conditions for the logistic map. In chapter 3 this property was made quantitative by introducing the (largest) Lyapunov exponent λ defined by (see equation 3.8),

$$\delta(t) = \delta(0) \exp(\lambda t). \tag{5.14}$$

Here $\delta(0)$ is the tiny separation between two initial points and $\delta(t)$ is the separation between the two trajectories emanating from these points after a time t. If in general the trajectories diverge exponentially—the signature of chaotic behaviour—λ will be found to be positive for almost all choices of the initial points. On the other hand if the trajectories diverge less rapidly than exponentially (for example, as t^n where $n > 0$), or if they run along side by side or move closer together, λ is zero or negative.

5.9

The largest Lyapunov exponent can be defined in an analogous manner for a map such as the one we have been considering. Computer calculations for r in the range $0 < r < r_\infty$ show that the Lyapunov exponent is zero or negative throughout this range. Thus the behaviour of the iterates is not chaotic. This is consistent with the fact that the 'motion' of the iterates (after an initial transient) is almost perfectly periodic at any value of r less than r_∞.

5.6 The logistic equation for 3.57 < r < 4.00

Once r exceeds $r_\infty = 3.5699456\ldots$, the iterates of the logistic map jump around chaotically in the interval $0 < x < 1$. That the behaviour is truly chaotic in the technical sense is confirmed by calculations of the Lyapunov exponent which show that λ is positive for almost all values of r in the range $r_\infty < r < 4.00$ (Schuster, 1984, p.34). Adding to the confusion, however, is the fact that within this chaotic régime there exist certain *windows of regular behaviour* for which $\lambda < 0$. These windows are characterized by periodic p-cycles with $p = 3, 5, 6, \ldots$ Each p-cycle displays a cascade of bifurcations occurring within a very narrow range of r-values. Thus the chaotic régime, covering the range $3.57 < r < 4.00$, is an extraordinary mixture of chaotic and regular behaviour. As our main interest in this chapter is in possible routes to chaos rather than in what happens when one gets there, we shall not pursue this any further.

5.7 The Feigenbaum constants α and δ

Our treatment of the logistic equation has emphasized the cascade of bifurcations and attendant period-doublings as the control parameter r goes from 3.00 to 3.57. We have seen in section 5.2 that exactly this kind of behaviour was found by Libchaber and Maurer (1980) in their experiments on Rayleigh-Bénard convection in liquid helium. Our discussion of the logistic equation would be incomplete, however, without mentioning the discoveries of M.J. Feigenbaum.

The main features of Feigenbaum's work are readily summarized. (For more details the reader is referred to the original papers (Feigenbaum, 1978, 1979) and the more accessible treatment in the book by Schuster (1984).) It is evident from Figure 5.6 that each time a bifurcation takes place the new attractive fixed points are distributed in the same way as before but on a reduced scale. Roughly speaking the width of each pitchfork is reduced by a factor of about 2.5 compared to its predecessor. For the sake of brevity we shall not give a precise mathematical definition of the width of a pitchfork after the n-th bifurcation, but we simply denote it as d_n. From high-precision numerical studies Feigenbaum found that the ratio of d_n to d_{n+1} approached a limiting value as n becomes large. This gives rise to the *universal constant* α defined as,

$$\alpha = \lim_{n\to\infty} \frac{d_n}{d_{n+1}} = 2.5029078750957\ldots \tag{5.1}$$

(The reason why it is called universal will be explained shortly.)

The positions r_n at which bifurcations occur (the bifurcation values of section 5.5) also show self-similarity on scaling. Roughly speaking, the separation between

r_{n+1} and r_n is about one fifth of the separation between r_n and r_{n-1}. Furthermore, as n increases, the ratio of these separations was found to approach a precise limiting value. This introduces a second universal constant denoted as δ (Feigenbaum, 1978):

$$\delta = \lim_{n\to\infty} \frac{r_n - r_{n-1}}{r_{n+1} - r_n} = 4.669201609\ldots \tag{5.16}$$

The existence of the constants α and δ in relation to the logistic equation is not completely unexpected. What *is* remarkable, however, is Feigenbaum's discovery that a whole class of maps, called quadratic maps, display a bifurcation cascade similar to that of the logistic equation and have exactly the same values of α and δ. This is the sense in which the constants are 'universal'.

The term *quadratic map* should be explained. It simply means that the mapping function $\tilde{f}(x)$ in the relation,

$$x \longmapsto r\tilde{f}(x), \tag{5.17}$$

has a *single* maximum in the interval $0 \le x \le 1$. When expanded in a Taylor series about the value x_0 at which the maximum occurs, $\tilde{f}(x)$ has the form (since $\tilde{f}'(x_0) = 0$),

$$\tilde{f}(x) = \tilde{f}(x_0) + \frac{1}{2}(x - x_0)^2 \tilde{f}''(x_0) + \ldots \tag{5.18}$$

The fact that the first term after the constant $\tilde{f}(x_0)$ on the right-hand side is quadratic in $(x - x_0)$, rather than cubic or some higher power, is a consequence of there being only a single maximum of $\tilde{f}(x)$ between 0 and 1. Examples of such maps, other than the logistic equation, are,

$$\tilde{f}(x) = r(1 - 2x^2), \tag{5.19}$$

and,

$$\tilde{f}(x) = rx(1 - x^2), \tag{5.20}$$

both of which were investigated numerically by Feigenbaum (1978). (Tabulated results in Appendix B of his paper show rapid convergence to the value δ given in (5.16).)

While acknowledging the universal character of α and δ for quadratic maps in the interval $0 \le x \le 1$, the reader may, nevertheless, be inclined to regard this class of maps as encompassing a good deal less than the universe. However, anyone familiar with the modern theory of critical phenomena (the physics of second-order phase transitions) will recognize the source of the hyperbole. In fact, the mathematical techniques developed by K.G. Wilson to describe scaling transformations and self-similarity near the critical temperature T_c (see, for example, Wilson and Kogut, 1974), are similar to Feigenbaum's methods for analying the scaling properties of quadratic maps (Feigenbaum, 1979). In the theory of critical phenomena the term 'universality class' is used to denote a class of phase transitions having the same critical exponents (which we shall not define). In an analogous manner, functions that give rise to a cascade of bifurcations which scale according to the same α and δ may be regarded as forming a universality class.

5.8 The intermittency route to chaos

Chaos has been shown to develop from regular behaviour in at least three distinctive ways, one of which is the period-doubling route which we have been concerned with up to this point. Another commonly occurring behavioural pattern is known as the *intermittency route to chaos.* It is illustrated by the visually-striking experimental measurements of Bergé, Dubois, Manneville and Pomeau (1980) shown in Figure 5.7.

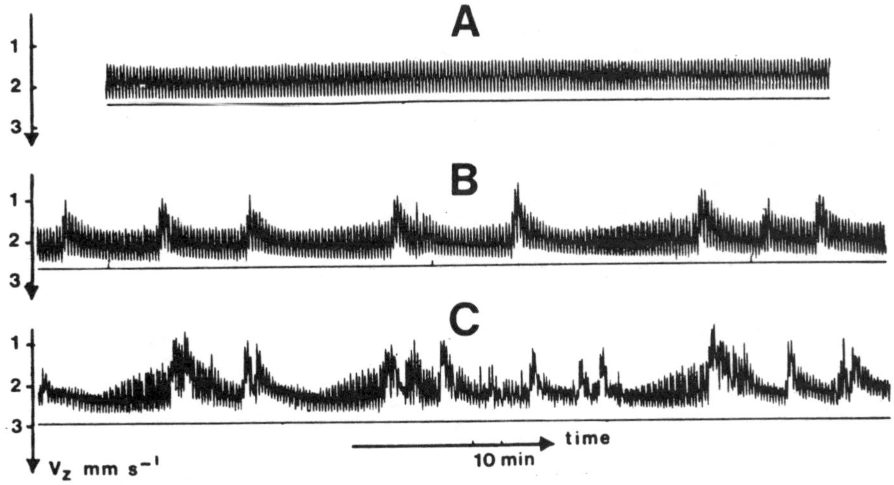

Figure 5.7 Time dependence of the vertical velocity component v_z measured at the centre of a Rayleigh-Bénard cell. A: $Ra = 270Ra_c$; B: $Ra = 300Ra_c$; C: $Ra = 335Ra_c$, where Ra_c is the critical Rayleigh number for the onset of convection in an infinite geometry. From Bergé, Dubois, Manneville and Pomeau (1980).

Bergé *et al.* studied the transition from steady convective flow to turbulent flow in the Rayleigh-Bénard configuration. The fluid was chosen to be silicone oil because of its relatively high Prandtl number ($Pr = 130$). The box in which it was confined measured 4.0cm by 2.4cm horizontally, with a depth of 2.0cm. Shown in Figure 5.7 are measurements, taken over several hours, of the vertical velocity component v_z of the fluid at the centre of the cell. The data were obtained using a laser technique at three different values of the Rayleigh number Ra. These are expressed in terms of the critical Rayleigh number $Ra_c = 1700$ for the onset of convection for fluid confined between two infinite plates. Figure 5.7A corresponds to about 4.5 hours of measurements for $Ra = 270Ra_c$. The velocity component oscillates in a regular fashion at a single frequency f_1 of about 0.1Hz. When the Rayleigh number was increased to $300Ra_c$, the behaviour shown in Figure 5.7B was obtained. Here the velocity component continues to oscillate at the frequency f_1, but from time to time the amplitude of the oscillation becomes strongly perturbed. After each perturbation, which Bergé *et al.* describe as a 'burst of turbulence', the behaviour relaxes back to the steady oscillatory behaviour of Figure 5.7A. At the still higher value of $Ra = 335Ra_c$ shown in Figure 5.7C. the turbulent bursts become more frequent and the so-called *laminar* intervals become shorter. The turbulent flow would be fully developed when the laminar intervals have essentially disappeared.

A simple mapping, which mimics a laminar phase interrupted by chaotic bursts, is sketched by Bergé *et al.* in their paper. It is discussed in greater detail in the book by Bergé, Pomeau and Vidal (1984) which includes an entire chapter on intermittency. The French researchers have identified three different types of intermittent behaviour, the one exemplified by Figure 5.7 being called type I. As a final remark, we note that the Lorenz equations discussed in chapter 3 have been found to display intermittency, changing from regular behaviour at $r = 166.0$ to intermittent bursts at $r = 166.1$.

Acknowledgment

I would like to thank Gerry Middleton for creating all but the first and last figures in this chapter. I am also grateful to him for many stimulating conversations about chaotic phenomena of all kinds, and for attempting to keep me up to date in the chaotic literature.

References

Bergé, P., Dubois, M., Manneville, P., and Pomeau, Y., 1980, Intermittency in Rayleigh-Bénard convection: Journal de Physique Lettres, v.41, p.L341-L345.

Bergé, P., Pomeau, Y., and Vidal, Ch., 1984, Order within chaos, Wiley, New York, and Hermann, Paris.

Feigenbaum, M.J., 1978, Quantitative universality for a class of nonlinear transformations: Journal of Statistical Physics, vol.19, p.25-52.

Feigenbaum, M.J., 1979, The universal metric properties of nonlinear transformations: Journal of Statistical Physics, vol.21, p.669-706.

Libchaber, A., and Maurer, J., 1982, A Rayleigh-Bénard experiment: helium in a small box, *in* Riste, T., ed., NATO Advanced Studies Institute on nonlinear phenomena at phase transitions and instabilities, p.259-286.

Malraison, B., Atten, P., Bergé, P., and Dubois, M., 1983, Dimension of strange attractors: an experimental determination for the chaotic régime of two convective systems: Journal de Physique Lettres, v.44, p.712-716.

Schuster, H.G., 1984, Deterministic chaos: an introduction, Physik-Verlag, Weinheim, Germany.

Wilson, K.G., and Kogut, J., 1974, The renormalization group and the ϵ expansion: Physics Reports, vol.12C, p.75-199.

1

CHAPTER 6

DIMENSIONALITY ANALYSIS OF OBJECTS AND SERIES DATA

Tony Fowler and Dan Roach*
* order of authors decided by lot
Department of Geology,
and Ottawa Carleton
Geoscience Centre
University of Ottawa
Ottawa, Ontario
K1N 6N5

6.1 Introduction

In this chapter we show how recent advances in applied mathematics, namely fractal geometry, and dimensionality analysis can be applied to geology. Mandelbrot (1977) has shown how various sets are manipulated in recursive fashion to yield highly complex tortuous curves. These curves are continuous but not everywhere differentiable, and hence were termed pathological or monstrous by earlier mathematicians. Although defined as "a set for which the Hausdorf-Besicovitch dimension strictly exceeds the topological dimension..." (Mandelbrot 1977) the precise definition of the word fractal is still evolving. Concepts such as self-similarity and scale-invariance are typically associated with fractals and Mandelbrot has recently termed fractal geometry as the geometry from Euclid to chaos.

Significant advances in the study of chaos and nonlinear dynamics have been made in the analysis of time series data. Many data sets appear to be composed of randomly distributed events, when analysed using traditional statistics. However, in some cases it is possible to use dimensionality analysis to determine if they were produced through deterministic or stochastic processes. The purpose of this chapter is to show how fractals can be used to quantify some textures, and to demonstrate new techniques of time series analysis.

6.2 Textures

6.2.1 Fractal Dimension

Fractals can be either exactly self-similar or statistically self-similar. Fractals such as the Von Koch snowflake curve generated by precise recursive geometrical procedures are exactly self-similar in that each small portion of the object, when suitably magnified, is exactly equal to a larger portion. Exactly self-similar fractals are mathematical entities not found in nature. Natural fractals, also known as random fractals, are statistically self-similar. Coastlines are well known examples of statistically self-similar fractals. Upon magnification, a small portion looks similar, but not identical, to a larger portion. They are fractal curves bounding areas having constant mass distributions (Fig.1a). On the other hand, fractal objects have distributions that are non-constant (Fig. 1b). The amount of material in the plagioclase crystal of Fig. 1a scales as the square of its radius in 2D, whereas the amount of material in the fractal object (Fig.1b) scales with radius to a smaller power, the fractal dimension.

6.2.2 Self-Affine Fractals

The Earth's surface, particularly in alpine regions, is often given as an example of a fractal. As with any surface, it's fractal dimension D_f is bounded between 2 and 3. When the earth's surface is viewed at close range, (i.e. from a ski-lift), there is definite sense of vertical relief. However, when viewed from a greater distance, (i.e. from a satellite), the vertical relief all but disappears. This apparent change in relief is due to different vertical and horizontal scaling. Objects possessing the property of non-uniform scaling, i.e. that scale different coordinates by different amounts, are known as self-affine fractals.

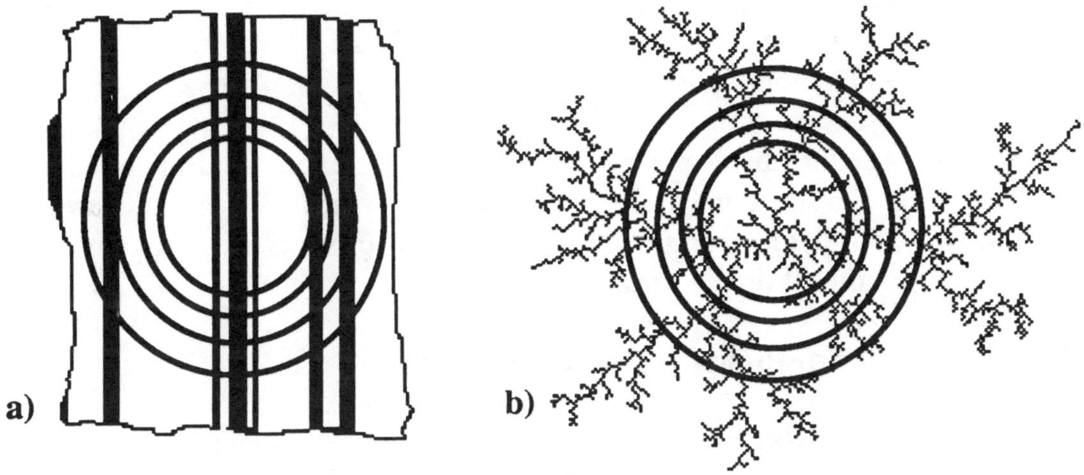

Fig. 1 In (a) the constant density plagioclase crystal is bounded by a fractal curve, and the mass in 2D scales as radius squared. b) The simulation of a far from equilibrium crystal texture having non-uniform distribution of mass. Here the 2d mass scales as a power of the fractal dimension. Note that the object is dilation symmetric i.e. composed of progressively larger void spaces. the cluster is a result of a computer simulation known as diffusion-limited aggregation (DLA). It is an excellent growth analogue of many natural patterns including crystals grown far from equilibrium (Fowler et al. 1989), patterns of fracture (Roach et al. 1991), dendritic stream channeling, lightning, and electrodeposition (e.g. Meakin and Fowler, 1991).

For measurement purposes, it is often necessary to convert self-affine fractals to self-similar fractals (see below). For self-affine surfaces such as the Earth's, this is accomplished by intersecting the surface with a horizontal plane, thus eliminating the vertical axis and leaving a family of self-similar curves or coastlines. This coastline set is referred to as a "zero-set". For example, if the earth's surface has a fractal dimension D_f=2.3, then any coastlines produced by intersection with a plane has a fractal dimension D_f = 1.3.

6.2.3 Measurement

There are many methods of estimating the fractal dimension of self-similar fractals. All of them are based on the relationship,

$$N=1/r^{Df} \qquad (6.1)$$

or taking logarithms of both sides,

$$D_f=\log(N)/\log(1/r) \qquad (6.2)$$

where N is the number of self-similar parts, each scaled by a factor of r from the whole comprising the fractal. Natural fractals are only fractal over a limited range of scale, as compared to an infinite range for mathematical fractals.

6.2.3.1 Mandelbrot-Richardson Plot

Consider a self-similar fractal curve, such as a coastline perimeter .This distance between the two end points is L_{max} . The curve can be measured using rulers of length (L), where

$$L = r * L_{max}, \quad \text{for } r<1 \qquad (6.3)$$

The measured perimeter (P) can be determined by

$$P = L * N \qquad (6.4)$$

where N is the number of steps of length (L). From eq. (6.1) we know that for fractals,

$$N = 1 / r^{Df}$$

Therefore, combining eqs 6.1, 6.3, and 6.4 yields

$$P = L * (L_{max} / L)^{Df}, \qquad (6.5)$$

or

$$P \sim 1 / L^{(Df-1)} \qquad (6.6)$$

Taking logarithms yields

$$\log(P) \sim -1 * (Df-1)* \log(L) \qquad (6.7)$$

If a range of ruler sizes is used then the fractal dimension can be estimated from the slope of a log(P) vs. log(L) where,

$$D_f = 1 - \text{slope} \qquad (6.8)$$

6.2.3.2 Box Dimension

The box method is analogous to the ruler method in that the fractal dimension Df is estimated by counting the number of squares (N) of side- length(L) containing an element of the set. From eq. 6.2, the fractal dimension D_f equals the slope of a log(N) vs. log(1/L) plot as measured over a range of box widths.

6.2.3.3 Mass Dimension

If each point in the fractal set is equivalent, (i.e. has the same mass), then one can measure the mass (M) of the portion of the fractal set contained within circles of radius (R), originating from a point. This measurement is repeated over a range of radii. Each point in the set is then used as the origin for measurement, and averages are taken. For fractal objects, the mass distribution scales as a non-integer power of radius

$$M \sim R^{Df}. \qquad (6.9)$$

A related technique, the correlation function plot computes the average density of material within concentric subshells (C(r)) with R, over all possible origins. For constant density objects, the slope of the log(C(r)) vs. log(R) plot is zero. For fractal objects the slope is negative and

$$D_f = 2 + \text{slope}. \qquad (6.10)$$

Consider sections of two Fe-Ti oxide crystals (Fig.2). Both are constant density items in that their mass distribution over any reasonable macroscopic scale of observation remains the same. They are bounded by curves (perimeters) that are dissimilar and might very well be termed euhedral and anhedral, or perhaps subhedral. Such qualitative descriptors of their boundaries are vague, however the crystal outlines can easily be quantified using fractal geometry. Figure 2 shows that the fractal dimension of the less tortuous boundary is 1.1 whereas the other is 1.3 . Clearly the box method is a very simple way of measuring crystallinity in an unambiguous manner. Although this measure does not characterize the crystal shape (i.e. equant or acicular) there are other standard techniques to measure two dimensional shapes.

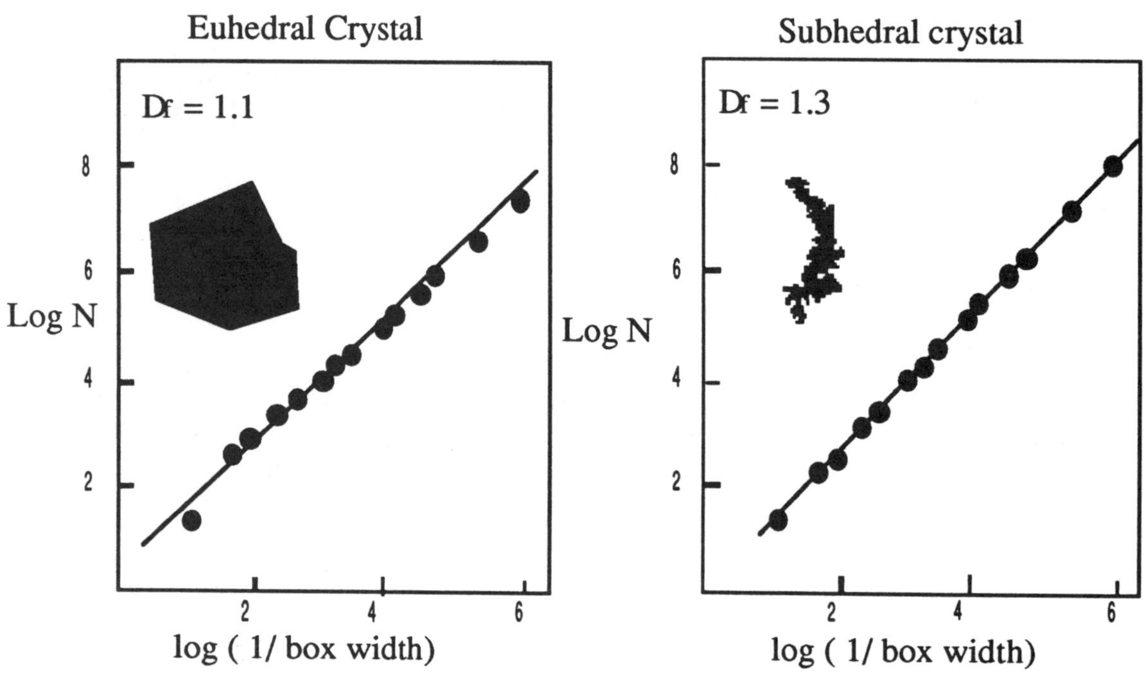

Fig.2 Box method plots for "euhedral" crystal and "anhedral" crystals. The slope of the plots gives the fractal dimension of the crystal perimeters shown inset.

6.2.3.4 D_f estimates for self-affine curves

Self-affine fractal curves do not scale equally in all directions. Consequently, they should not be measured directly using the methods listed above (Voss 1988). Instead, they should be reduced to their self-similar zero-sets. The fractal dimension of the curve's zero-set D_z is calculated using any of the above techniques, and Df of the original curve equals $D_z + 1$.

6.2.3.5 Slit Island Method

The box method can be extended to higher topological dimensions to estimate the fractal dimension of surfaces. Applying equation (6.1) to surfaces, N becomes the number of cubes (or spheres), of size L needed to cover the surface. For real surfaces this method is often difficult to implement. Furthermore, if the surface is self-affine, estimates of D_f made using the box method may be erroneous. Therefore, it is easier and more reliable to construct and measure self-similar zero-sets of the fractal surface.

The slit island method was introduced by Mandelbrot et al. (1984) in their study of fracture surfaces in metal. Zero-sets were created by intersecting the surface with planes parallel to the main fracture plane. The zero-sets emerged as islands having fractal coastlines. The measured perimeter of the coastlines depends on the length of the ruler (L) and the fractal dimension D_f, as in eq. (6.5), which reduces to

$$P \sim L_{max}^{\ Df} \qquad (6.11)$$

when using a constant ruler size ($L << L_{max}$). The area (A) of the emerging islands varies according to

$$A \sim L_{max}^2, \qquad (6.12)$$

or

$$A^{1/2} \sim L_{max} \qquad (6.13)$$

Combining equations (6.11) and (6.13) yields

$$P \sim A^{Df/2} \qquad (6.14)$$

And taking logarithms,

$$\log(P) \sim (D_f / 2) * \log(A) \qquad (6.15)$$

Therefore, the slope of a log(P) vs. log(A) plot for islands of various sizes yields ($D_f/2$) .

Joint surfaces in rock often have irregular radiating plumose-type surface features (Fig. 3).

Fig.3 This self-affine fractal, a plumose joint was simulated by Roach et al. 1991. The patterns are commonly found on the joint surfaces of fine grained rocks. These are self-affine, as evident from their rough appearance at fine scales and planar appearance at large scales. Their fractal dimension has been estimated using the slit island method (Roach et al., 1991). In this case, a zero-set was created by intersecting the fracture surface with a series of planes parallel to the plane of fracture. The resulting log (P) vs. log(A) plot (Fig.4) for islands yields a slope of 0.63 . Therefore $D_f = 1.26$.

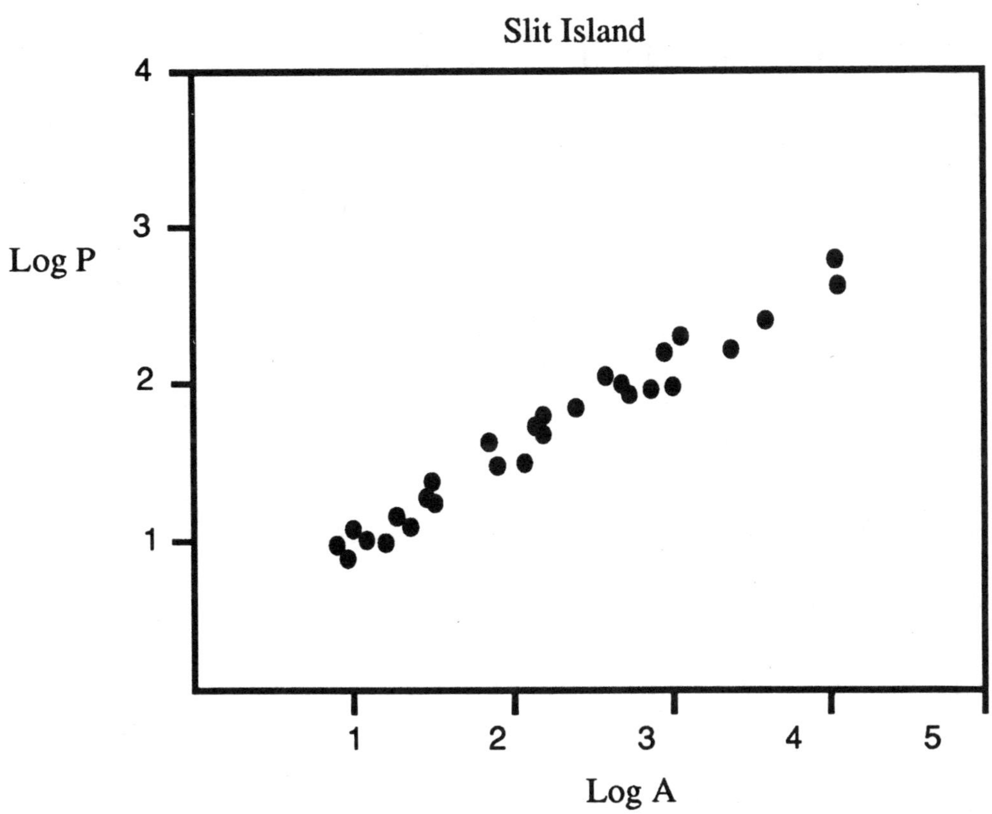

Fig. 4 Log perimeter versus Log area plot for zero-sets of a plumose joint. Slope = 0.63 and D_f = 1.26

6.2.4 Computer Algorithms

6.2.4.1 Box Method Algorithm

The box method is the easiest to program because it utilizes the square nature of the grid imposed by the digitizing process. The box method simply involves counting the number of boxes (N) of width (L) that contain a piece of the fractal set. To achieve this, a counting grid is imposed on the image. Grid widths are systematically varied from 1 pixel to the maximum size (in pixels) of the fractal set. For each grid width (L), the number of boxes (N) that contain a piece of the image are counted and stored. The location of the fractal set relative to the counting grid is varied to achieve average values for N. A typical box method program has the following structure,

load image

LOOP1: relocate image relative to counting grid

```
        LOOP2: set new grid width (L)
            LOOP3: count boxes (N) containing piece of image
          ENDL3;
      ENDL2;
   ENDL1;
    determine Df = slope of log(N) vs. log(L) plot
```

6.2.4.2 Mass Dimension (Correlation function) Algorithm

Assuming each pixel is equivalent to one unit of mass, this algorithm calculates the mass of the fractal (M) contained within a circle of radius (R) centered about a point on the fractal. To estimate the fractal dimension, the mass is calculated as the radius is systematically varied from 1 pixel to half the maximum image size. Average values for M are achieved by using every point in the fractal set as the center of the circle. A typical algorithm follows;

```
    load image
     LOOP1: select new center point for circle
        LOOP2: set new radius (R) for circle
            LOOP3: calculate total mass (M) within circle
          ENDL3;
      ENDL2;
   ENDL1;
    determine Df = slope of log(M) vs. log(R) plot
```

6.2.4.3 Slit Island Algorithm

The slit island method requires measurements of area (A) and perimeter (P) for individual islands in the fractal set. The area is simply the total number of pixels making up the island. The perimeter is equal to the total number of pixel edges constituting the shoreline of the island. A typical algorithm follows;

```
load image
  LOOP1: select island for measuring
      LOOP2: count total number of pixels = area
             count shoreline pixel edges = perimeter
    ENDL2;
ENDL1;
 determine Df = 2 x slope of log(P) vs. log(A)
```

6.3 Time Series Analysis

Time series or sequences of events can be analysed using modifications of some of the techniques shown above. In addition, series of spatially varying scalar data (e.g. tree rings, where the spatial variation records a time variation) can be analysed by the methods. One might very well pose the question what possible use this could be? We can answer this by looking at classes of time sequences. There are those that are completely deterministic and characterized by resonance. Common examples are unambiguous and familiar, such as night and day, the change of seasons, vibrations of a violin string. Thanks to Newtonian mechanics we can usefully predict the season, daytime or nighttime, given a time in the future anywhere on the planet. Other sequences are characterized by what might be termed randomness, chaos, or dissonance. Examples are the records of earthquakes, weather patterns, the results of coin tosses, and much modern classical music. Some of these may be governed by truly stochastic processes, others by truly deterministic processes. Until recently traditional statistical methods of analysis failed to yield any new insight into the governing processes. Large numbers of observations or populations were studied in order to understand processes better through statistical or thermodynamic approaches. Now we are able to show that some of these chaotic outputs are in fact governed by relatively simple dynamical systems. Although we comprehend the dynamics of some systems more concretely our predictive powers are only little improved over those of probabalistic approaches. However, this is useful because at the very least we know if the dynamics are deterministic, and the limitations in the ranges of our predictions.

In geology the time scales of processes are so long compared to recorded human history, and the precision to which we can measure events is poor. Hence it is difficult at first sight to see any potential for time series applications other than to short period phenomena such as the record of earthquakes, geysers, and volcanic eruptions. However records of previous events preserved in rocks are thought to be the signature of time-varying processes and we can use a proxy for time e.g. bed or zone thickness for rhythmic cycles of sedimentation or mineral patterning.

Table 1: N = observation number; Logout = values for the logistic equation @ C=3.999, Logdif = differences of successive values; Ranout = values of random numbers; Radif = differences of successive values; Lorout = x- values Lorenz system @ constants in eqs. for x,y,z = 10, 28 & 2.66 respectively; Lordif = differences of successive values;

N	Logout	Logdif	Ranout	Randif	Lorout	Lordif
1	.4598		28		9.77	
2	.9930	.5332	69	41	6.52	-3.25
3	.0277	-.9653	84	15	-6.84	13.36
4	.1078	.0801	94	10	-6.9	-.06
5	.3846	.2768	62	-32	-10.24	-3.34
6	.9462	.5616	67	5	-9.56	.68
7	.2034	-.7428	86	19	1.98	11.54
8	.6478	.4444	24	-62	8.39	6.41
9	.9122	.2644	63	39	-10.24	18.63
10	.3203	-.5919	38	-25	-8.73	1.51
11	.8704	.5501	6	-32	-2.61	6.12
12	.4511	-.4193	86	80	-10.73	-8.12
13	.9899	5388	54	-32	11.43	22.16
14	.0401	-.9498	99	45	5.28	-6.15

15	.1537	.1136	0	-99	-1.41	-6.69
16	.52	.3663	65	65	-12.7	11.29
17	.9979	.4779	26	-39	8.82	21.52
18	.0086	-.9893	94	68	-4.31	13.13
19	.034	.0254	2	-92	-3.15	1.16
20	.1312	.0972	82	80	9.61	12.76
21	.4557	.3245	90	8	-8.73	18.34
22	.9916	.5359	23	-67	-9.2	-.47
23	.0332	-.9584	7	-16	-6.81	2.39
24	.1283	.0951	35	28	-5.24	1.57
25	.4471	.3188	30	-5	-13.79	-8.55
26	.9883	.5412	58	28	4.51	18.3
27	.0464	-.9419	21	-37	-5.36	-9.87
28	.1768	.1304	46	25	-7.72	-2.36
29	.5819	.4051	6	-40	-11.11	-3.39
30	.9726	.3907	72	66	-7.56	3.55
31	.1064	-.8662	17	-55	.97	8.53
32	.3801	.2737	10	-7	16.93	15.96
33	.942	.5619	94	84	-9	25.93
34	.2184	-.7236	25	-69	-8.82	.18
35	.6825	.4641	21	-4	-6.94	1.88
36	.8663	1838	31	10	-6.46	.48
37	.4632	-.4031	75	44	-12.88	-6.42
38	.9940	.5308	96	21	-1.16	11.72
39	.0237	-.9703	63	-33	-10.13	-8.97
40	.0926	.0689	43	-20	2.68	12.81
41	.3359	.2433	36	-7	11.59	8.91
42	.8918	.5559	82	46	-11.52	23.11
43	.3857	-.5061	69	-13	-4.23	7.29
44	.9472	.5615	65	-4	1.4	5.63
45	.1999	-.7473	51	-14	11.04	9.64
46	.6394	.4395	18	-33	2.17	-8.87
47	.9217	.2823	37	19	-4.23	-6.4
48	.2884	-.6333	88	51	-4.65	-.42
49	.8204	.5320	61	-27	-1.49	3.16
50	.5891	-.2313	38	-23	14.97	16.46

6.3.1 Dimensionality

This term causes difficulty to many people because of the fact that we can only visualize three dimensions yet frequently we encounter more in the scientific literature. We can conceptualize more by thinking in parameter space, not our usual space where one can of course only accommodate three mutually perpendicular axes. Thus space-time is a four dimensional "space" that uniquely defines the temporal-geographic relationships of anything we wish. We can add other variables as we please. For instance inclusion of the term granite to the above would be a five dimensional system representing the distribution of granites through time. As geologists we would look at these data as a series of time slices (maps) and reduce the dimension to four because we do not generally care about the present day elevation distribution of granite. In computers dimensionality is the dataspace or data array containing variables describing the system. In the example above these are {lat., long., elev., age, granite (y/n)}. In dynamical systems dimensionality represents the number of independent variables, or degrees of freedom of the system.

6.3.2 Difference plot

Table 1 has data output arrays from the logistic equation $X_{n+1} = CX_n (X_{n-1})$ for fifty iterations (n=50), with C= 3.999 and a seed value X_n = 0.5. This equation is made up of one constant, and two variables, one dependent and one independent. Therefore, as a dynamical system it has dimension one (1). Table 1 also has a listing of data from tables of random numbers (from Fisher and Yates, in Kennedy and Neville, 1976, Tbl. A-14). Figure 5 shows the graph of the two data series. Note that there is no qualitative difference between the two. They both appear devoid of information, i.e. uniform or random.

A straight forward approach to distinguishing between deterministic and stochastic behaviour used by Farmer in his now famous dripping tap experiment is shown if Fig. 6 Here the data for successive values have been subtracted (Table 1) and plotted in sequence $(X_{n+1} - X_n)$ vs. $(X_{n+2} - X_{n+1})$; $(X_{n+2} - X_{n+1})$ vs. $(X_{n+3} - X_{n+2})$ etc. One immediately notices that the disorder of the data in the random sequence is preserved whereas that of the logistic has a qualitatively different behaviour. Note that the plot relates increments of change of successive values, and that the logistic changes in a smooth fashion. This is good empirical evidence that it is deterministic, i.e. future values depend upon past values in some easily predictable manner. Clearly this is not so for the data from the random number tables.

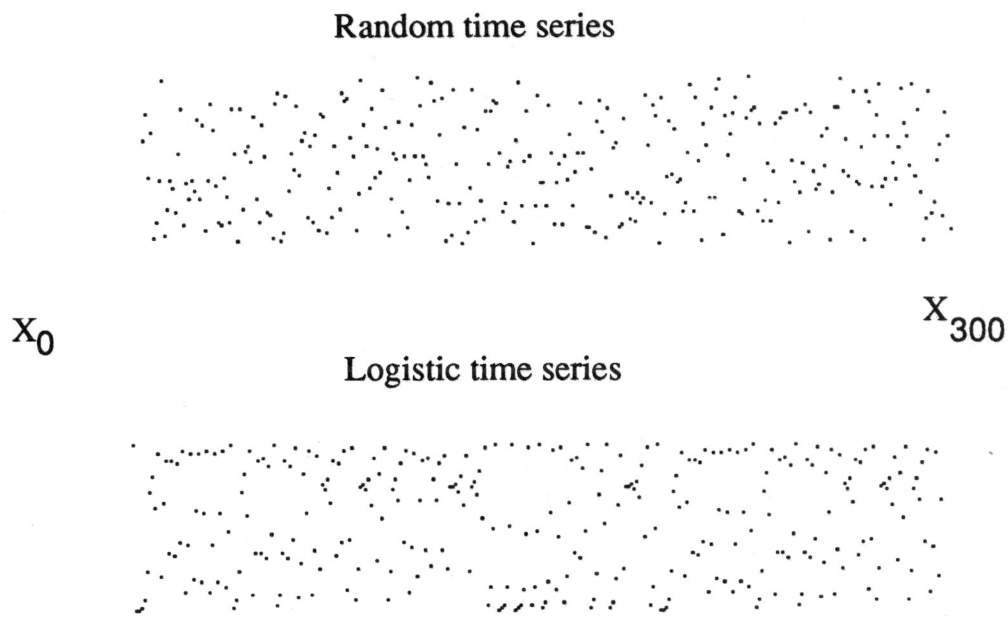

Fig. 5 Graphical display of time series. Top random numbers, bottom logistic equation. Values are plotted on the ordinate, time on the abcissa for 300 observations.

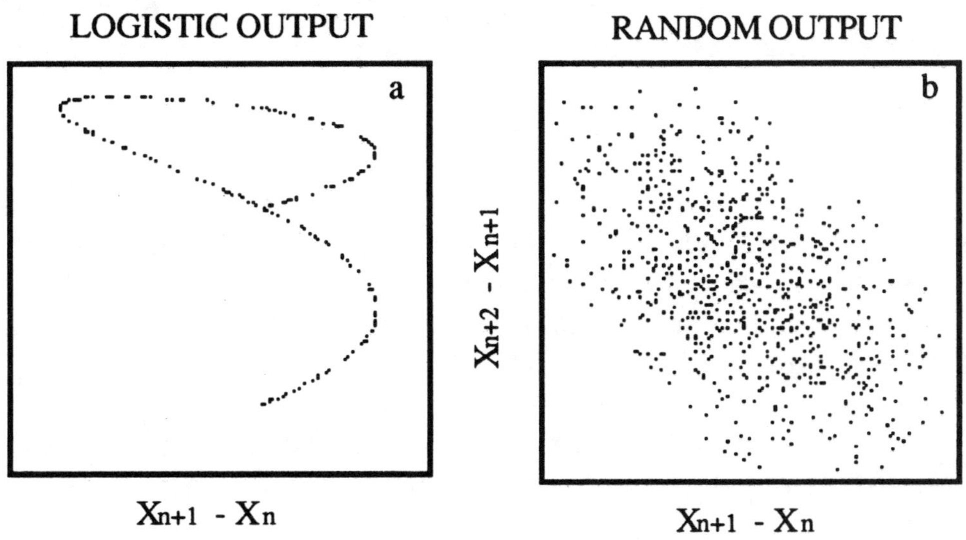

Fig. 6a Difference plot of the logistic equation; C= 3.999, note continuity of curve. **b** Difference plot of random numbers, no continuity is evident.

6.3.3 Zero order

A robust measure of multi-dimension system behaviour is provided by the zero-order technique of (Farmer & Sidorowich 1987). The method predicts future values in the time series by examining old values. It relies on the fact that deterministic data are spatially restricted and sequenced by the system i.e. they inhabit an attractor, whereas stochastic data are uniformly distributed. The technique can be easily demonstrated in one dimension using the data from the logistic output (Table1). We can predict the value at time (n=2) using the value from (n=1) 0.4598 and simply searching through the data set for the value closest to it, in this case 0.4632 (n= 37). The value at (n=38) 0.9940 becomes our predictor. Note that it is very close to and successfully predicts the value at (n=2) of 0.9930. A similar process for the randomly distributed data of table 1 has the value of 28 at (n=1). Its closest value in the time sequence is 26 at (n=17) leading to a poor predicted value of 94 at (n=18) whereas the actual value at (n=2) is 69.

Instead of thinking of the values in the time series as being scalar variables we can think of their time evolution as being the displacement of some physical quantity i.e. a vector. Therefore the zero order technique is extended into higher dimensions by transforming the time series into dimensional vectors. For instance the data of the Lorenz system are transformed into a two dimensional array in the following manner $\{X_n, X_{n+1}\}$; $\{X_{n+1}, X_{n+2}\}$, a three dimensional array $\{X_n, X_{n+1}, X_{n+2}\}$; $\{X_{n+1}, X_{n+2}, X_{n+3}\}$ and so on. These are resolved (square root of the sum of the squares) and predictions of the next value in the time series are then made by searching for the nearest neighbour values in the resolved vector space in a manner similar to the one dimensional case. Correlations between the predicted and actual values can be made as in Figs.7 and 8 which show very low correlations for random data as compared to data from the Lorenz system. An estimate of dimension can be made from a graph of the correlation coefficient versus the dimensionality E.

Figure 7 shows that for data from the Lorenz attractor the dimension lies between >1 and <5, that is where the correlations are most significant (Sugihara and May (1990). This is the so-called embedding dimension (E) and is equivalent to a minimum system dimension of 2 by the expression of Takens 1981.

$$E_{min} < 2D_f + 1 \qquad 6.16$$

Figures 7 and 8 also show the results of changing the sampling interval. One expects coarsening the interval to result in the deterioration of the correlations calculated from a dynamical system, whereas the random data should show no systematic difference. This result arises because of the fact that as we skip data points in a deterministic system the precision to which we can make

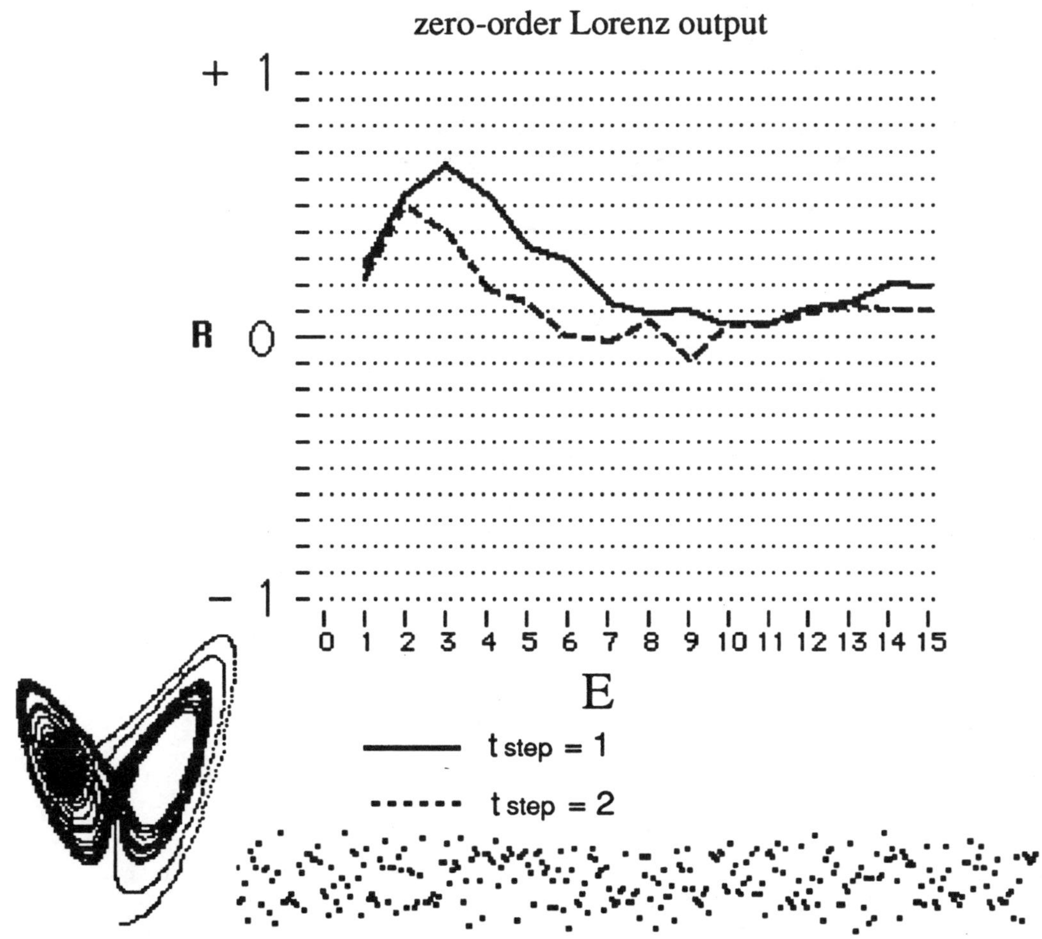

Fig.7 Zero-order analysis for one of the variables from the Lorenz system. R is the correlation coefficient between predicted and actual variables (N = 100 observations). E = the embedding dimension. Most significant correlation is at E=3. The plot has two curves t_{step}= 1 which uses every data point, and t_{step}= 2 which uses every second

data point. Correlations deteriorate as the time step is lengthened. Projection of the Lorenz attractor, bottom left, and data series for the analysed coordinate appears across the bottom of the figure. See table one for the first half of these data.

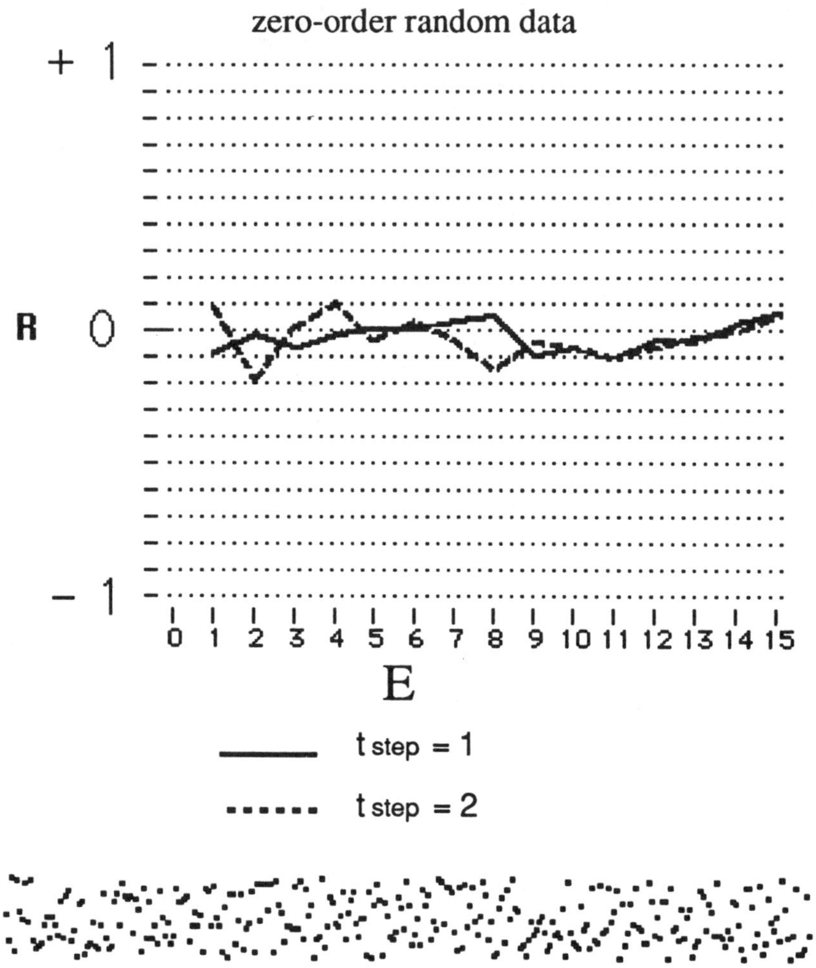

Fig.8 Zero-order analysis for random data. R is the correlation coefficient between predicted and actual variables (N = 100 observations). E = the embedding dimension. Note that there are no significant correlations. The plot has two curves t_{step}= 1 which uses every data point, and t_{step}= 2 which uses every second data point. Correlations do not deteriorate as the time step is lengthened. The data series appears across the bottom of the figure. See table one for the first half of these data.

predictions must decrease, whereas in a random system no predictions are possible at the outset hence the correlations remain poor as the sampling interval is changed.

6.3.4 Simplex

The "simplex" method of Sugihara and May (1990) is a more robust technique for the analysis of time series. Here again the data are transformed into a set of E-dimensional vectors. However, rather than a single value or vector to make a prediction the simplex uses several nearby values by monitoring their time evolution, i.e. their trajectories. This has an advantage over the zero-order method in that the prediction is not purely based on on the evolution of a single similar value. Obviously the difficult part in nonlinear prediction is making extrapolations where the system is most nonlinear, that is where the attractor trajectory is changing direction quickly. Thus the use of several values allows a rapid change in trajectory to be "sensed" whereas a single value prediction may not "sense" or "reach" an impending rapid change in trajectory ahead of the predictee.

The simplex is the smallest "E-volume" that contains the predictee in E-space. The E+1 vertices (vectors) of the simplex are chosen as the nearest valued neigbours. These are advanced an appropriate time step forward into the future of the time series e. g. their next scalar values are picked for a single time step. The predictee's change in position with respect to the simplex is mapped by "giving exponential weight to the original distances from the relevant neighbours" (Sugihara and May 1990). We interpret this to mean that exponential weight is given to the nearest values, chosen in the previous step to make the prediction. Although we are unaware of any proof that the exponential weighting is in fact correct, it makes sense at an intuitive level due to the fact that trajectories diverge at an exponential rate, hence those closest to the predictee likely predict better than remote ones. Certainly Sugihara and May (1990) show that the technique works very well on natural data from biological systems and overcomes the requirement of huge data sets. We show a comparison of the zero-order and the simplex in Fig.9. Both graphs plot the predicted versus the actual value, hence perfect prediction has a correlation coefficient, and slope of one (1). Both graphs show data for the same 20 values of the logistic equation for

C=3.99. Note that the simplex has significantly higher correlations and a slope closer to one than the zero-order. The real advantage of the technique appears to be that significant correlations can be made on relatively small (n ~ 100's) data sets.

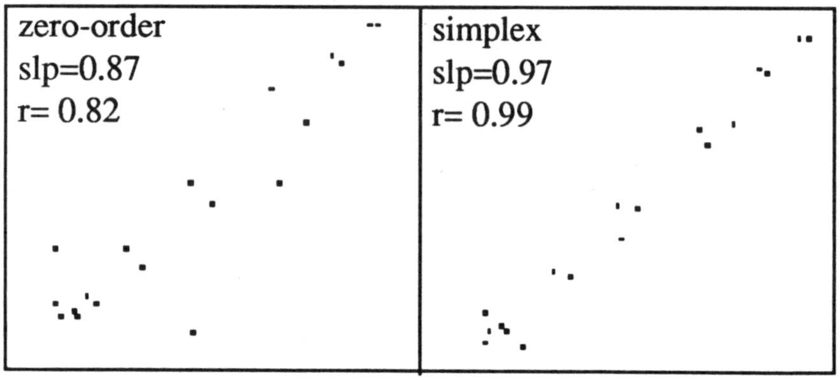

Fig. 9 Results of zero-order analysis of a time series composed of 20 values from the logistic equation.

6.3.5 Spatial Correlation Function

This technique (see Goodings earlier chapter, this volume) works in the same way as the mass dimension above in that it computes for increments of radius R, the number (C(r)) of E-dimensional points (resultant vectors) whose magnitude is less than R. Plots of log C(r) vs. log r are constructed for all E dimensions. Because $C(r) \sim R^D$ the slope of the lines generated converge to a value of D despite increasing E. In contrast, random data have no such power law relationship and as a consequence show no convergence of slopes (Fig. 10). Unlike the above methods that rely on making predictions a step at a time into the future this algorithm sorts all the data with respect to R in each E-dimensional volume. Therefore the method suffers in that it requires very large data sets n>5,000 and is sensitive to noise. Our experiments show that little (~5%) sampling noise in output of the logistic equation made the results indistinguishable from those of stochastic sets, whereas the zero order and simplex easily saw through this masking. The large data set requirement and the sensitivity to "noise" restrict the use of the technique for geological data sets.

6.3.6 Algorithms

6.3.6.1 zero - order

load time series

LOOP1: transform time series to E-dimensional vector space

LOOP 2: get E-dimensional vector for prediction, save next scalar value as predictee

LOOP 3: find nearest E-dimensional vector, save next scalar value as predictor

ENDL3;

ENDL2;

ENDL1;
correlate predictee vs. predictor
Plot correlation coefficient R vs. E embedding dimension.

6.3.6.2 Simplex

load time series

LOOP1: transform time series to E-dimensional vector space

LOOP 2: get E-dimensional vector for prediction, save
next next scalar value predictee

LOOP 3 form simplex from E+1 nearest values. Save next scalar values, form predictor giving exp. weight to nearest neighbours

ENDL3;

ENDL2;

ENDL1;
correlate predictee vs. predictor
Plot correlation coefficient R vs. E embedding dimension.

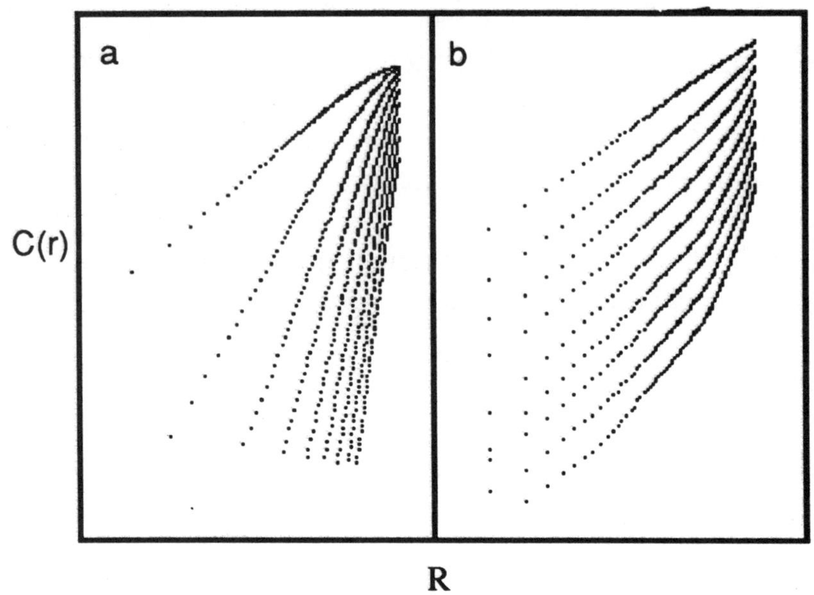

Fig. 10 Correlation function plot a) output from the logistic equation and b) random data, 5,000 values each. Note the continuous change in slope for the analysis of the random data whereas the data of the logistic equation define a constant slope for small R over 5 dimensions.

6.4 References

Farmer, J. D., and, Sidorowich, J.J. 1987: predicting chaotic time series. Phys. Rev. Lett. 59, 845-

Fowler, A.D., Stanley, H.E. and Daccord, G. 1989: Disequilibrium silicate mineral textures: fractal and non-fractal features. Nature 341, 134-138.

Kennedy, J.B., Neville, A.M. 1976: Basic statistical methods for engineers and scientists. Harper & Row, New York, 489p.

Mandelbrot, B.B. 1977: The fractal Geometry of Nature. W.H. Freeman & Sons, San Francisco, 468p.

Mandelbrot, B.B., Passoja, D.E. and Paullay, A.J. 1984: Fractal character of fracture surfaces of metals. Nature 308, 721-722.

Meakin, P., and Fowler, A.D. 1991: Diffusion-limited aggregation in the earth sciences. Submitted for publication in American Association of Petroleum Geologists, Special Publication, Fractals in the earth sciences.

Roach, D. E., Fowler, A.D., and Fyson, W.K., 1991: Shatter cones and plumose joints: Fracture growth simulation and fracture surface analyses. Submitted to Geology.

Sugihara, G., and May, R.M. 1990: Nonlinear forecasting as a way of distinguishing chaos from measurement error in time series. Nature, 344, 734- 741.

Takens, F. 1981: Detecting strange attractors in turbulence. Lecture notes in mathematics 898, ed. D. A. Rand. L. 'S. Young. Springer Verlag, 390 p.

Voss, Richard F., 1988, Fractals in nature: From characterization to simulation. Chapter 1 in H-O. Peitgen and D. Saupe, eds., The Science of Fractal Images. N.Y., Springer-Verlag, p.21-70.

CHAPTER 7

GEOCHEMICAL SELF–ORGANIZATION: AN APPLICATION OF NONLINEAR DYNAMICS

J. Chadam
Department of Mathematics and Statistics
McMaster University
Hamilton, Ontario
L8S 4K1

7.1 Introduction

The use of nonlinear dynamics to understand self–organization phenomena has been an extremely successful approach in chemistry, physics, biology, etc. (Nicolis and Prigogine, 1977). It turns out that many phenomena of geological interest (shape instabilities in solidification, reactive flows in porous media, oscillatory zoning in crystals, diagenetic bedding, metamorphic differentiation, etc.) are tractable by these methods (Ortoleva, 1991a,b). In this note we shall describe from this viewpoint one important geochemical phenomenon, reactive flow in porous media. Specifically we shall be interested in how the shapes of the evolving reaction fronts are selected by the nonlinear dynamics.

In this note *self–organization* means the spontaneous transition of a non–equilibrium system from a non–patterned state to a patterned state without the intervention of a patterned external cause. Some of the "physical" requirements for self–organization to occur are: disequilibrium; a destabilizing feedback (coupling) mechanism competing with a restabilizing mechanism; and noise (fluctuations).

The essential characteristic of the resulting mathematical model is that it be non–linear.

We begin with a brief introduction to the ideas of bifurcation and stability theory and how they are useful as methods for the study of self–organization. We then model reactive flow in porous media as reaction–transport equations. The interpretation of these governing equations is given in terms of the geochemical processes rather than abstractly. One finds that in the limit of weakly soluble solids the reaction zone (where both reaction and transport are significant and outside of which transport is dominant) shrinks to an infinitesimally thin moving interface. This is the correct geological problem to solve because we are interested in the shape of the reaction zone rather than the complicated behavior occurring within it. In mathematical terms we obtain a Stefan–like moving free boundary problem for the *unknown* reaction interface. That is we are faced with the problem of solving a set of partial differential equations not in the classical framework with fixed boundaries but rather with one of the boundaries being an unknown and on which an additional condition is given which governs its evolution. The reaction–transport equations and the more specialized Stefan problem are the nonlinear dynamics governing the evolution of the reaction front.

In spite of the fact that these equations and problems are complicated, it is important that the reader see them in order to appreciate that interesting problems in geochemistry often lead to very difficult mathematical problems which require the most modern techniques for their resolution. Finally, using bifurcation and stability theory on the Stefan problem and numerical simulation of the reaction–transport equations we study the morphological stability of the reaction interface. Our objective is to study whether initially high–symmetry fronts evolve as such or whether they self–organize into fronts with patterned shapes or fingers, and whether they experience tip splitting, budding, meandering etc.

7.2 Mathematical Preliminaries

One of the mathematical techniques used in analytically treating self–organization is bifurcation and stability theory. The most simple (and at the same time most general) place to begin is with an evolution equation for some unknown quantity u;

$$u_t = N(u,\lambda). \tag{7.1}$$

As suggested in the previous section, N is nonlinear in u and λ is a parameter–eventually the bifurcation parameter. For the present discussion the only important feature of the nonlinearity is to allow for the time–independent problem,

$$N(u, \lambda) = 0, \tag{7.2}$$

and hence the evolution equation (7.1), to have multiple time–independent solutions for each λ. While it is usually the case that equation (7.1) along with the initial data.

$$u(t_0) = u_0(x) \tag{7.3}$$

has a unique solution (i.e. (7.1) and (7.2) is a deterministic problem) a more interesting question is to determine to which of the time–independent solutions of (7.2) does this solution of (7.1) and (7.3) tend. Since it is also usually impossible to explicitly determine the solution of (7.1) and (7.3) one must resort to other methods–bifurcation and stability.

The basic idea to determine whether nearby solutions are attracted to a given time–independent solution, $\bar{u}$, is the following. Suppose initially a solution of (7.1), u, is nearby to $\bar{u}$; i.e. $u(t_0) - \bar{u} \simeq 0$. Consider the difference at a later time, $z(t) = u(t) - \bar{u}$. It satisfies

$$z_t = u_t = N(u, \lambda) = N(u, \lambda) - N(\bar{u}, \lambda). \tag{7.4}$$

since $\bar{u}_t = 0$ and $N(\bar{u}, \lambda) = 0$. Now if t is only slightly larger than t_0, then one expects $u(t) - \bar{u} = z(t)$ to still be small but with the important question being whether it is larger or smaller than the initial separation $z(t_0)$. Just as in Calculus (the Mean Value Theorem or Taylor's series) the right hand side of (7.4) can be approximated by

$$N(u, \lambda) - N(\bar{u}, \lambda) \cong N_u(\bar{u}, \lambda)\,(u - \bar{u}) = N_u(\bar{u}, \lambda)z, \tag{7.5}$$

where N_u is the (functional) derivative of N with respect to u. It is completely determined by the form of N and the chosen time–independent solution $\bar{u}$. Thus z, the deviation from $\bar{u}$, satisfies the linear equation

$$z_t = L_\lambda z. \tag{7.6}$$

where $L_\lambda = N_u(\bar{u}, \lambda)$. Typically, N and hence L_λ is a differential operator in a spatial variable (eg. involving $\partial^2/\partial y^2$, say). Since (7.6) is a linear problem one can expand in the eigenvectors of L_λ, $\{\zeta_m\}$, and take Laplace transforms in t, or equivalently set

$$z_m(t) = e^{\sigma_m t}\, \zeta_m \tag{7.7}$$

to obtain

$$\sigma_m\, \zeta_m = L_\lambda\, \zeta_m. \tag{7.8}$$

Thus $\sigma_m = \sigma_m(\lambda)$ is a λ–dependent eigenvalue of L_λ. If the spectrum of L_λ, $\{\sigma_m\}_{m=1}^{\infty}$, has negative real parts then clearly from (7.7) all modes $z_m(t)$ are shrinking and so u is tending toward $\bar{u}$. That is, $\bar{u}$ is linearly stable (since the above analysis deals with the linearized equation (7.6)). On the other hand, if even one eigenvalue has positive real part, then the corresponding mode in increasing, forcing u away from $\bar{u}$. Typically this occurs by having the real part of this eigenvalue go from negative to positive as the parameter λ crosses a critical value λ_c. Bifurcation theory continues then to show that the above ideas carry over to the original nonlinear equation (7.1) in a sufficiently small neighborhood of the bifurcation point $(\lambda_c, \bar{u})$.

Finally, to relate this mathematical method to self–organization, suppose $\bar{u}$ represents the evolving unpatterned state. If $\bar{u}$ is stable, then although random fluctuations might cause the spontaneous appearance of some mode z_m, this perturbation will subsequently decay. On the other hand, if $\bar{u}$ is unstable with respect to some mode, the random fluctuation will eventually cause it to appear and hence drive the evolution away from $\bar{u}$ to another patterned state which is stable. As the parameter λ increases there may be several modes to which $\bar{u}$ is unstable. The actual patterned state which emerges involves a competition between these modes through the nonlinear equation (7.1) with the remaining modes decaying to zero. Typically these large amplitude considerations (i.e. (7.4) is not valid) require numerical simulations of (7.1). To this end, the linearized stability analysis described here assists in locating the parameter values where interesting behavior occurs.

7.3 The Geological Phenomenon.

Consider an aquifer consisting of an insoluble porous matrix (e.g. quartz sandstone) with some soluble mineral (e.g. calcite) partially filling the pores. If water is forced through this porous medium, the soluble component will be dissolved out upstream and the water will become saturated sufficiently far downstream. Between these extremes there is a dissolution zone across which the soluble mineral content – and hence the porosity and permeability – change from their original downstream values to their final, altered values upstream. The question of interest is whether patterned shapes of this dissolution zone can arise out of a homogeneous environment. Notice that if a bump (in the surfaces of equal porosity values) in the reaction zone exists at some time, the flow of the unsaturated waters tends to be focused to the tip of the bump via Darcy's law since inside the bump (on the upstream side) the permeability is greater than in the neighboring regions (see Fig. 1). Thus dissolution is enhanced at the tip causing it to advance more rapidly. This is the porosity (equivalently, permeability since we shall assume a phenomenological relation between them) change/flow destabilizing feedback mechanism (Chadam, Ortoleva et al, 1986, 1987a, 1987b, 1987c, 1988). On the other hand, diffusion from the sides of the tip raises the concentration of the solute in the water which is focusing at the tip and hence will decelerate this advancement. The competition between these two processes will determine, through the nonlinear dynamics whether the bump will decay to a high–symmetry, unpatterned zone, restabilize to a morphologically more complicated dissolution zone or possibly, completely destabilize to a fractal–like surface. The disequilibrium occurs because the inlet waters are not in equilibrium with the unaltered, homogeneous porous medium. Since fluctuations in the porous medium will certainly be present, we have shown that the three basic requirements for self–organization are present in this situation.

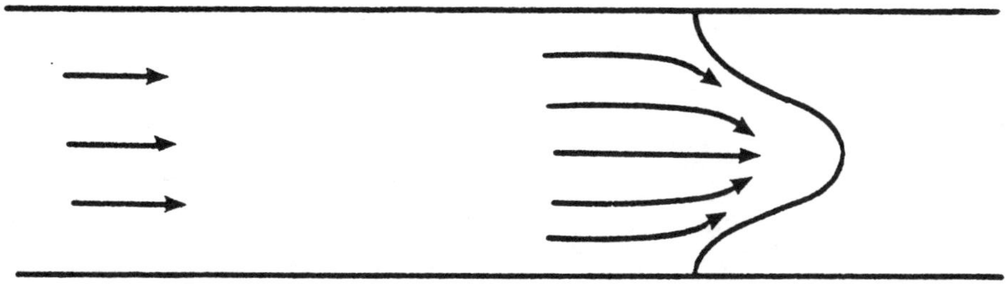

Fig. 1. Focusing of flow to tip of porosity level curve.

This phenomenon appears to be quite prevalent causing scalloped geological reaction fronts to emerge in a wide range of wavelengths (5 cms in siltstone near Bloomington, Indiana as in Fig. 2, to 10 kms in uranium roll type deposits in South Texas).

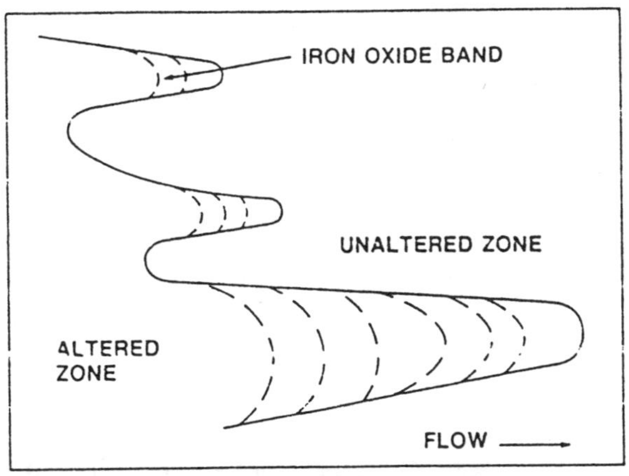

Fig. 2. Trace from a photograph of a loose block of the Bordon Group, Southern Indiana containing a scalloped reaction front. The wavelength of the scallops is 5 to 6 cm. Because of the iron oxide bands indicated by the dashed lines, the water is inferred to have flowed to the right.

This shape selection (self–organization) process certainly is important in many geochemical situations (e.g. the diagenesis and evolution of mineral deposits, oil and gas reservoirs, the dynamics of breakout from nuclear and chemical waste repositories, in situ coal gasification, enhanced oil recovery, etc.).

7.4 The Mathematical Models.

We shall describe two mathematical models of this phenomenon, the first a coupled set of reaction–transport equations which, by taking a certain distinguished limit, reduces to the second, a moving free boundary problem.

The rate of increase of porosity, φ, (equivalently, the rate of dissolution of the soluble mineral) is proportional to the reaction rate:

$$\varphi_t = -k(\varphi_f - \varphi)^{2/3}(c - c_{eq}), \tag{7.9a}$$

where k is the reaction rate constant, φ_f is the final porosity after complete dissolution and c is the concentration of solute in water with its equilibrium concentration being c_{eq}. The 2/3–power indicates that we are considering surface reactions. The solute concentration per rock volume, φc, satisfies a mass conservation equation:

$$(\varphi c)_t = \nabla\cdot[\varphi D(\varphi)\nabla c + \varphi c\, \kappa(\varphi)\, \nabla p] + \rho\, \varphi_t, \tag{7.9b}$$

where D and κ are the porosity dependent diffusion and permeability and ρ is the density of the mineral being dissolved. In (7.9b) p is the pressure and Darcy's law for the velocity $\underline{v}$ has been used in the form (the viscosity is taken to be constant)

$$\underline{v} = -\kappa(\varphi)\, \nabla p. \tag{7.10}$$

Finally the conservation of water implies

$$(\varphi\rho_w)_t + \nabla\cdot(\varphi\rho_w\underline{v}) = 0, \tag{7.11}$$

which with the constancy of the density of water, ρ_w, and Darcy's law implies

$$\varphi_t = \nabla\cdot(\varphi\kappa(\varphi)\, \nabla p). \tag{7.9c}$$

The three equations (7.9a,b,c) are to be solved for the three unknowns φ,c,p subject to the imposed asymptotic conditions,

$$c \to 0,\ \varphi \to \varphi_f \text{ and } p_x \to p_f' \text{ as } x \to -\infty \tag{7.9d}$$

and

$$c \to c_{eq},\ \varphi \to \varphi_0 \text{ and } p_x \to ? \text{ as } x \to +\infty \tag{7.9e}$$

and initial data. The conditions (7.9d) say that at the inlet $(x = -\infty)$, the water is free of solute, the porous medium has reached its final altered state in which all the soluble mineral has been previously dissolved out and a horizontal pressure gradient (equivalently velocity, through Darcy's law) has been imposed. Similarly, the conditions (7.9e) say that at the outlet far downstream $(x = +\infty)$, the water is saturated with solute, the porous medium is in its original state and the pressure gradient is to be determined as part of the solution. We take zero flux boundary conditions on the transverse boundaries of the aquifer. Problem (7.9) is a system of reaction–transport equations which plays the role of the nonlinear dynamics governing the evolution of the system.

One might suspect that a planar travelling wave solution would be the base high–symmetry solution. Because (7.9b and c) are in divergence form, an integration from $-\infty$ to $+\infty$ gives algebraic equations which are soluble for the velocity of the travelling wave and the unknown $p_x(+\infty, t)$. On the other hand it appears impossible to obtain the profiles of φ,c,p explicitly. Indeed, in the simpler case when there is no change in the porosity $(\varphi_f = \varphi_0)$ the existence of these profiles could proven abstractly but the actual shapes could not be computed (Chadam, Ortoleva et al, 1986). This situation is quite unsatisfactory since, to do a stability analysis, one requires the base planar solution explicitly. Fortunately there is a way out of the predicament since in most geological examples of interest the (transverse) size of the reaction front is several orders of magnitude larger than the thickness of the reaction zone and the details inside this zone are not of interest per se except in the manner in which the cumulative effect provides the dynamics for the evolution of the reaction front on the larger scale. Mathematically, this means scaling problem (7.9) with the objective of finding a parameter which, in typical geological situations, is small. To this end, (Chadam, Ortoleva et al, 1987 a,b,c). Let

$$\epsilon = c_{eq}/\rho, \tag{7.12}$$

and examine the limit $\epsilon \longrightarrow 0$ (typical values of ϵ range from 10^{-3} to 10^{-10}). One expects the thickness of the resulting reaction front separating the two values of φ to be very thin (indeed, it is $O(\epsilon^{1/2})$) and the front itself to move very slowly. Thus we introduce a slow dimensionless time $\tilde{t} = \epsilon(k\ c_{eq})t$ and other inessential scalings for the space variables, concentration, pressure, diffusion and pressure.

Taking the so–called large solid density limit, $c_{eq}/\rho \longrightarrow 0$, using matched asymptotics (Chadam, Ortoleva et al, 1987c, 1988) we arrive at the following moving free boundary problem.

If the dissolution interface is taken to be located at x = R(y,t) with R *unknown*, the equations upstream are:

$$\left.\begin{aligned} &\nabla^2\gamma + \nabla\gamma\cdot\nabla p = 0, && (7.13a)\\ &\varphi = \varphi_f, && (7.13b)\\ &\nabla^2 p = 0, && (7.13c)\end{aligned}\right\} \quad \text{in } x < R(y,t),\quad 0 < y < \pi$$

while downstream they are

$$\left.\begin{array}{l}\gamma \equiv 1, \\ \\ \varphi = \varphi_0 \\ \\ \nabla^2 p = 0,\end{array}\right\} \quad \begin{array}{l}\text{in } x > R(y,t) \\ 0 < y < \pi\end{array} \qquad \begin{array}{r}(7.14a)\\ \\ (7.14b)\\ \\ (7.14c)\end{array}$$

Notice that the porosity assumes only two values, the final value φ_f after all the mineral has been dissolved out and the initial value φ_0 with all the mineral present. Here we have purposely taken φ_0 constant to show that patterned fronts will occur even when the initial porous medium is homogeneous.

Across the dissolution interface (essentially defining it) the porosity changes discontinuously from its unaltered value φ_0 to its final value $\varphi_f > \varphi_0$. Using matched asymptotics (Chadam et al., 1987c), the other boundary conditions on the unknown moving dissolution interface follow from continuity of γ and p, and conservation of water and solute

$$\gamma = 1, \tag{7.15a}$$

$$p_- = p_+, \tag{7.15b}$$

$$\left[\frac{\partial p}{\partial x} - \frac{\partial p}{\partial y} \cdot \frac{\partial R}{\partial y}\right]_- = \Gamma\left[\frac{\partial p}{\partial x} - \frac{\partial p}{\partial y} \cdot \frac{\partial R}{\partial y}\right]_+, \qquad \begin{array}{l}\text{on } x = R(y,t) \\ 0 < y < \pi\end{array} \tag{7.15c}$$

$$\frac{\partial \gamma}{\partial x} - \frac{\partial \gamma}{\partial y} \cdot \frac{\partial R}{\partial y} = R_t, \tag{7.15d}$$

where $0 \leq \Gamma = \varphi_0 \kappa_0 / \varphi_f \kappa_f < 1$ is a measure of the porosity change ($\kappa_0 = \kappa(\varphi_0)$ and $\kappa_f = \kappa(\varphi_f)$ are permeabilities). Furthermore, subscripts + or – indicate evaluation just up– and down–stream from the reaction front. Equation (7.15d) relates the rate of advancement of the moving dissolution interface to the flux of the concentration and is called a Stefan condition.

To complete the problem we add the asymptotic conditions

$$\gamma \to 0, \frac{\partial p}{\partial x} \to -\nu_f \qquad \text{as } x \to -\infty, \tag{7.16a}$$

$$\gamma \to 1, \frac{\partial p}{\partial x} \to -\nu_0 \qquad \text{as } x \to +\infty, \tag{7.16b}$$

and the zero–flux conditions for γ and p on the aquifer walls,

$$\frac{\partial \gamma}{\partial y} = 0, \; \frac{\partial p}{\partial y} = 0 \;\text{ for }\; y = 0, \pi, \quad -\infty < x < \infty. \tag{7.17}$$

Here the imposed pressure gradient (equivalently imposed velocity via Darcy's law) at $x = -\infty$ is, in the present dimensionless variables, $\nu_f = v_f \, L \, / \, D_f \pi$ where $v_f = -\kappa(\varphi_f) \, \nabla p$ is the imposed (unscaled) velocity at $x = -\infty$, $D_f = D(\varphi_f)$ is the diffusion coefficient in the altered medium and L is the original transverse dimension of the aquifer. The similar pressure gradient at $x = +\infty$, ν_0, is to be determined. It is important to notice that here, as with the porous medium, there is no imposed pattern so that if any develops, it is the result of self–organization.

7.5 Shape Stability of the Reaction Front.

In what follows, we shall obtain the planar, constant velocity solution of equations (7.13a – 7.17). The linearized stability of this solution then gives a precise value of the parameter ν_f (in terms of Γ) for which the planar solution loses stability to another, more structured, solution. In the language of bifurcation theory we determine the critical parameter value for which the spectrum of the linearized problem changes sign from negative to positive, thus determining the location of a possible "bifurcation point" beyond which nonplanar reaction front morphologies may exist. Numerical simulations in the vicinity of this critical value of ν_f give an indication of the patterned front which emerges.

One can easily check (Chadam, Ortoleva et al. 1987c, 1988) that the following, planar solution (denoted with an over–bar) solves problem (7.13a – 7.17):

$$\overline{R}(t) = \nabla t, \quad V = \nu_f = v_f L \, / \, D_f \pi, \tag{7.18}$$

$$\overline{\gamma}(x,t) = \begin{cases} e^{-\nu_f(x-\nabla t)} & x < \nabla t, \quad (7.19a) \\ 1 & x > \nabla t, \quad (7.19b) \end{cases}$$

$$\overline{p}(x,t) = \begin{cases} -\nu_f(x-\nabla t) & x < \nabla t, \quad (7.20a) \\ -\nu_0(x-\nabla t) & x > \nabla t, \quad (7.20b) \end{cases}$$

where $\nu_0 = \varphi_f \, \nu_f \, / \, \varphi_0$. In order to examine the stability of the above planar solution with respect to bumps, we study the linearized problem obtained for small ($\delta << 1$) perturbations of the form

$$R(y,t) = \bar{V}t + \delta r_m(t)\cos(my), \tag{7.21a}$$

$$\gamma(x,y,t) = \bar{\gamma}(x,t) + \delta\, \gamma_m(x) r_m(t)\cos(my), \tag{7.21b}$$

$$p(x,y,t) = \bar{p}(x,t) + \delta\, p_m(x)\, r_m(t)\cos(my). \tag{7.21c}$$

These equations can be solved explicitly (Chadam, Ortoleva et. al., 1987c) for γ_m and p_m and inserting into the Stefan condition gives the following condition on the amplitude $r_m(t)$ of the cosmy bump (prime denoting t–derivative):

$$r_m'(t) = \frac{\nu_f}{1+\Gamma}\left[\nu_f - \left[\nu_f^2 + 4m^2\right]^{1/2} + (1-\Gamma)\,|m|\right] r_m(t). \tag{7.22}$$

This differential equation indicates that the cos(my) is linearly stable or unstable depending on the sign of the coefficient. In bifurcation terminology, writing $r_m(t) = e^{\sigma(m)t}$, the spectrum of the linearized problem, $\sigma(m)$, can be obtained from the following dispersion relation:

$$\sigma(m) = \frac{\nu_f}{1+\Gamma}\left[\nu_f - \left[\nu_f + 4m^2\right]^{1/2} + (1-\Gamma)\,|m|\right] \tag{7.23}$$

the graph of which is given in Fig. 3.

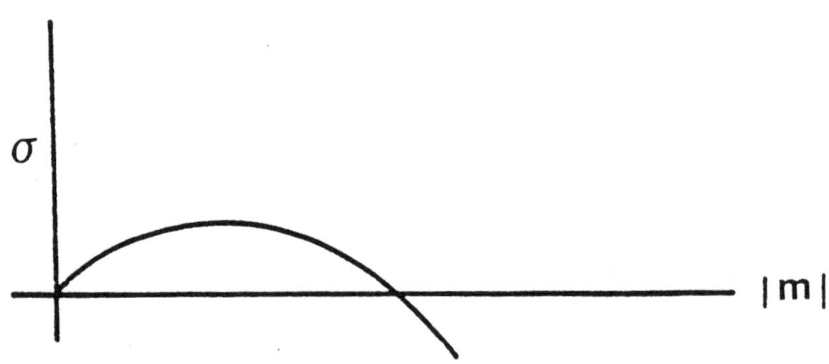

Fig. 3. Graph of dispersion relation (7.23)

Because $\Gamma < 1$, the planar solution is unstable to long wavelength perturbations and stable to short wavelengths. The critical wave number ($|m_0|$ at which $\sigma(m_0) = 0$) is given by:

$$|m_0| = \frac{2(1-\Gamma)}{(3-\Gamma)(1+\Gamma)}\,\nu_f. \tag{7.24}$$

Since our channel width has been normalized to π, the first mode which can be carried is $|m_0| = 1$ giving, from (7.24), the critical value of the bifurcation parameter ($\nu_f = v_f L / D_f \pi$),

$$\nu_c = \nu_c(\Gamma) = \frac{(3-\Gamma)(1+\Gamma)}{2(1-\Gamma)}. \tag{7.25}$$

This verifies analytically that shape instabilities can arise in the reaction interface associated with this reaction–infiltration mechanism in a homogeneous environment. Moreover, as is expected physically, this analysis shows that larger flow speeds, larger transverse dimensions and larger porosity/permeability changes promote the instability, while larger diffusion coefficients inhibit the instability (diffusion is stabilizing, as mentioned earlier). The limit $\Gamma \to 1$ (i.e. no porosity change) suggests that it is very difficult to produce instabilities. Indeed, the stability of planar fronts when $\Gamma = 1$ has been verified in a separate analysis (Chadam, Ortoleva et al, 1986). Thus this reaction–infiltration instability can occur only if "significant" amounts of soluble mineral are dissolved causing a porosity change.

As the parameter $\nu_f = v_f L/D_f \pi$ increases it passes the critical value

$$\nu_{c,m} = \frac{(3-\Gamma)(1+\Gamma)}{2(1-\Gamma)} |m|, \qquad m = 1, 2, \ldots, \tag{7.26}$$

for the loss of stability of the planar solution to the various modes, cosmy, $m = 1, 2, \ldots$. We can show analytically, using a weakly non–linear stability analysis (Chadam, Ortoleva et al., 1988), that, in the neighborhood of these bifurcation points the cos(my) solutions are indeed the restabilized solutions of the planar front. Thus we have established a bifurcation diagram consisting of a sequence of pitchforks at these critical values as depicted in Fig. 4.

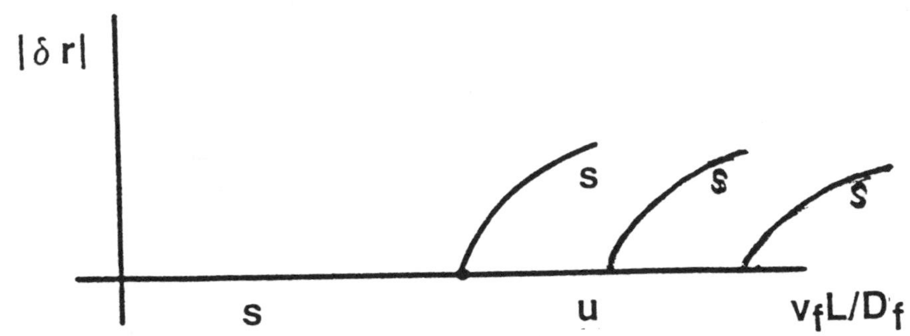

Fig. 4. Bifurcation diagram for equations (7.13a – 7.17).

On the other hand, if we are in a range of parameter values where many modes are unstable, they interact through the nonlinear dynamics (7.13–17) in a very complicated way. It is precisely this competition of the linearly unstable modes with the short wavelength stable modes which spontaneously (autonomously) generates the shape of the evolving interaction front. This aspect of the problem has been studied numerically and the results are summarized in the next section.

7.6 Numerical Simulations.

The free boundary problem (7.13a – 7.17) which was so useful for the analytical treatment is unsuitable for a numerical treatment since it would require complicated front–tracking methods. Instead we use the reaction–transport model (7.9) with ϵ small (but not so small that the reaction front is excessively thin giving rise to large gradients and stiff numerical problems). In the figures below the position of the front is approximated by the porosity level curve $\varphi = (\varphi_f + \varphi_0)/2$ whose position is shown at successive time intervals. At time zero the domain is uniformly cemented except for a small high porosity "bump" located near the inlet on the left.

In Fig. 5a one sees that for $\nu_f \ll \nu_c = \nu_{c,1}$ (i.e., all modes are linearly stable) the perturbation dies out leaving a stable evolving planar front. In Fig. 5b where $\nu_f \simeq \nu_c$, the planar front eventually loses stability to a constant velocity cos(y) front. In Fig. 5c where $\nu_f \simeq 5\nu_c$, the many unstable modes couple through the nonlinear dynamics to produce an elongating finger.

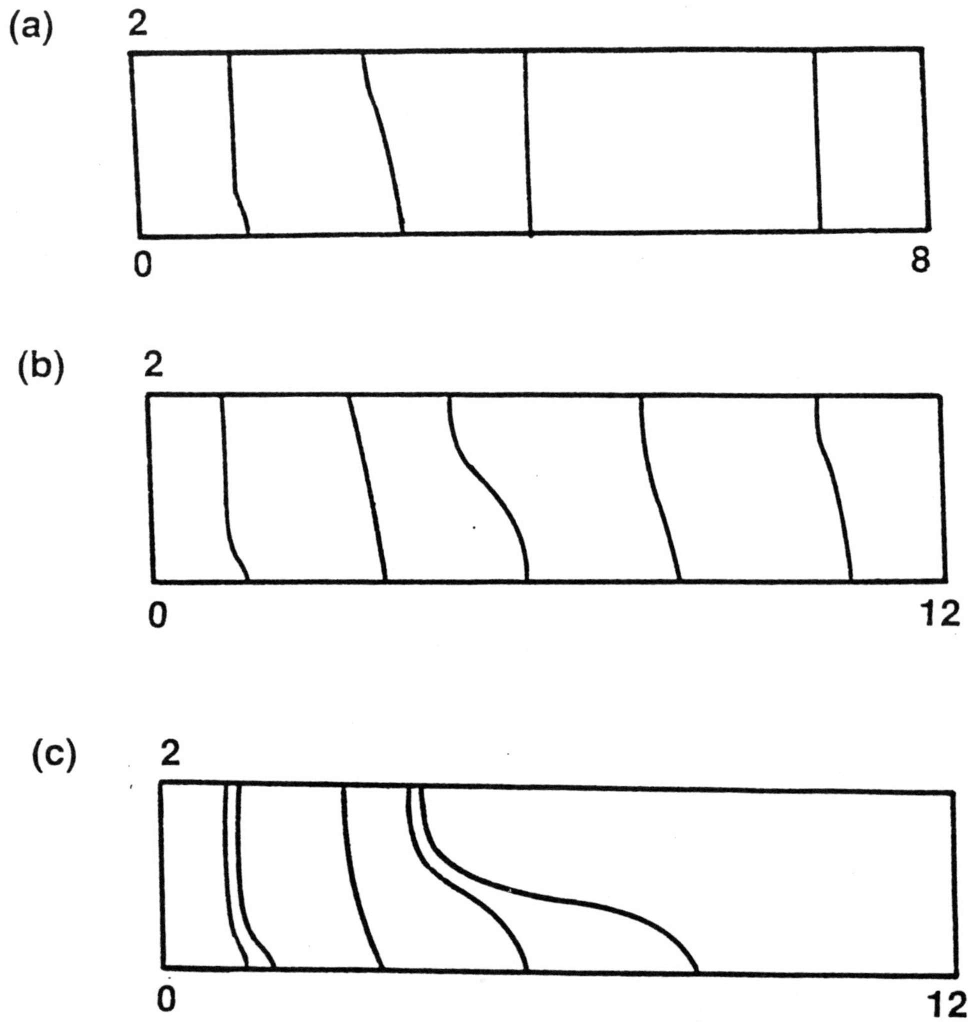

Fig. 5. Numerical simulations of the evolution of an initial bump in the reaction zone when (a) $\nu_f \ll \nu_c$, (b) $\nu_f \simeq \nu_c$ and (c) $\nu_f \gg \nu_c$.

Higher precision calculations were performed at Indiana University (Chen and Ortoleva, 1990) allowing for much longer time evolution of the fronts and displaying new dynamic self–organization. In Fig. 6a one sees that for faster inlet flows or wider domains, the advancing finger can destabilize causing "tip splitting". For different values of the parameters, the central finger can dominate the side fingers giving rise to "budding" as in Fig. 6b. The buds arise periodically in time in a frame moving with the advancing central finger. This suggests that a critical value of the bifurcation parameter exists at which the steady advancing nonplanar state (finger) undergoes a (secondary) Hopf bifurcation to a temporally oscillatory budding state. With nonsymmetric data as in Fig. 6c, this process may change the course of the main branch leading to the "meandering" of the dissolution channel.

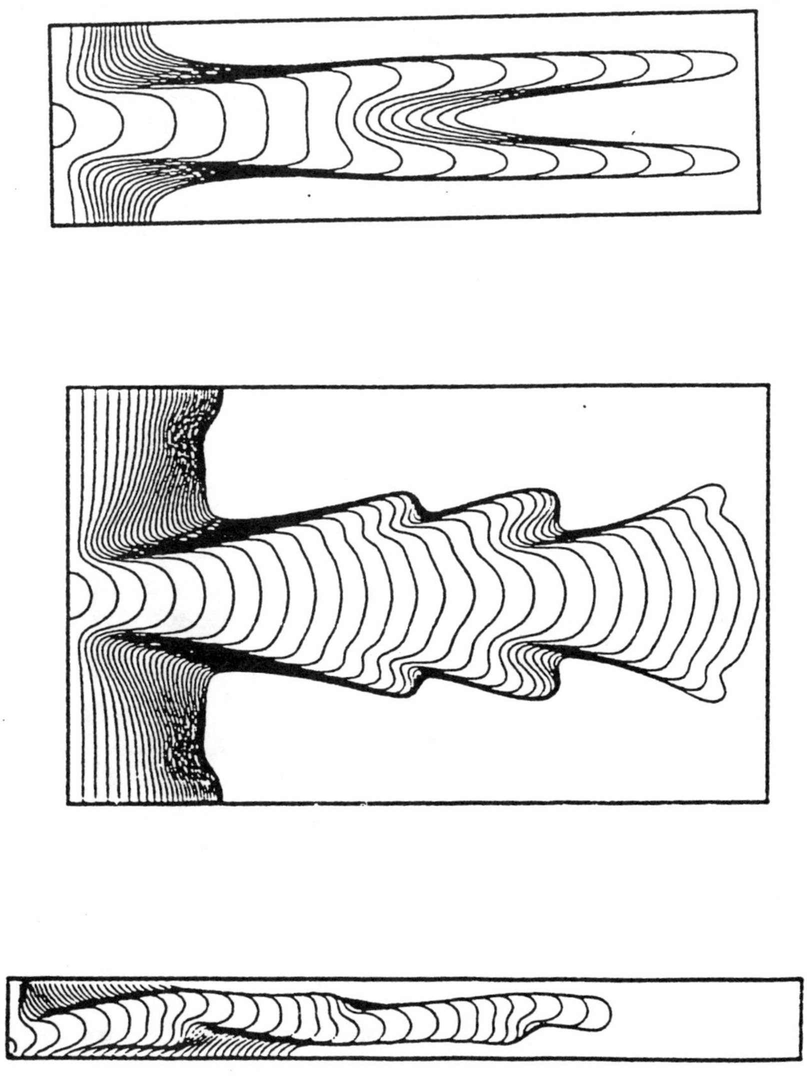

Fig. 6. Numerical simulations exhibiting a) tip splitting, b) budding and c) meandering (from Chen and Ortoleva, 1990).

7.7 Summary.

In this note we study reactive flows in porous media and, in particular, the shape stability of the evolving, thin reactive zones. The key ingredients of a destabilizing feedback mechanism (porosity–change/flow instability) completing with a restabilization process (diffusion) in a non–equilbrium situation containing noise suggests that the problem can be viewed as a self–organization phenomenon. The problem is modeled mathematically in general as a system of nonlinear reaction–transport equations and in the special case of geological interest (the large solid density limit) as a Stefan–like moving free boundary problem. The latter is amenable to the methods of nonlinear dynamics (bifurcation and stability theory) to analytically obtain the critical values of the bifurcation parameter (the Peclet number) for which planar reaction fronts lose stability to more structured travelling fronts. To study the interaction of many unstable modes competing through the nonlinear dynamics, the original reaction–transport equations are simulated numerically under a variety of initial conditions and parameter values. In addition to showing the development of the initial finger, more precise calculations (Chen and Ortoleva, 1990) show subsequent development (secondary bifurcations) of "tip splitting", oscillatory "budding" and the resulting "meandering" of the main channel.

These fronts support a rich class of steady and dynamic structures which can be understood qualitatively from the viewpoint of self–organization phenomena and obtained analytically using the methods of nonlinear dynamics.

References

Chadam J., Auchmuty G., Merino E., Ortoleva P. and Ripley E., (1986). The structure and stability of moving redox fronts, SIAM J. Appl. Math. 46: 588–604.

Chadam J., Ortoleva P., Merino E. and Moore C., (1987a). Geochemical Self–Organization I: Feedback mechanisms and modeling approach, Amer. J. Sci. 287: 979–1007.

Chadam J., Ortoleva P., Merino E. and Moore C., (1987b). Self–Organization in water–rock interaction systems, II: The reactive–infiltrate instability, Amer. J. Sci. 287: 1008–1040.

Chadam J., Ortoleva P. and Sen A., (1987c). Reactive infiltration instabilities, IMA J. of Appl. Math. 36: 207–220.

Chadam J., Ortoleva P. and Sen A., (1988). Weakly nonlinear stability of reaction–percolation interfaces, SIAM J. of Appl. Math. 48: 1362–1378.

Chen, W. and Ortoleva, P., (1990). Reaction front fingering in carbonate–cemented sandstones, Earth–Sci. Rev. 29: 183–198.

Nicolis, G. and Prigogine, I. (1977). Self–Organization in Non–Equilibrium Systems, Wiley, New York.

Ortoleva, P., (1991a). Geochemical Self–Organization, Oxford University Press, New York.

Ortoleva P., (1991b). Nonlinear Chemical Waves, Manchester University Press, Manchester.

CHAPTER 8

MAGMATIC PHENOMENOLOGY AS NONLINEAR DYNAMICS: ANTHOLOGY OF SOME RELEVANT EXPERIMENTS AND PORTRAITS

Herbert R. Shaw
U. S. Geological Survey
Menlo Park, CA 94025

8.1 Introduction

I was tempted to entitle this Chapter, "Count Rumford, Circle Maps, and Magma", because these names symbolize phenomena that have several things in common, namely dissipation, recursive behavior (reiterative and/or cyclic processes), feedback, oscillatory coupling (including numerical, mechanical, and thermal, as the cases may be), self-organization, and universality, to name what I consider to be an essential if not exhaustive set of attributes appropriate to natural systems. At the same time, these three names broadly identify how this Chapter is organized: (1) I will describe -- and illustrate with an experiment -- some effects I consider to be physically important ("rheologically" relevant in the broadest sense) pertaining to thermal dissipation of mechanical energy in a form analogous to that of Rumford's experiments; (2) Stereotypical nonlinear models will be used to illustrate some of the commonly expected regimes of behavior in dissipative systems including those that may be approximately conservative in the spatial sense: circle maps are especially instructive for this purpose, but Hamiltonian systems per se are not discussed; and (3) Throughout, I attempt to point toward some of the ways these ideas combine to govern aspects of magmatic regimes (actual or possible) in planetary or cosmological contexts, with examples drawn from my experiences in studying the evolution and rheology of magmatic systems.

Consideration of these topics shows that they generally can not be discussed effectively without considering their analogies, parallels, and identities. For example, certain types of physical experiments effectively act as analog or even digital computers in terms of illustrating recursive capability and some relative numerical regimes. Without attempting to be comprehensive, electronic devices (besides the computer itself) are well known in this regard. The review by Van Buskirk and Jeffries (1985) of the nonlinear oscillatory regimes of semiconductor devices illustrates a rich array of nonlinear dynamical phenomena appropriate to my present emphasis on interactive dissipative transport processes. A study of electronic transport by Gwinn and Westervelt (1987) precisely mimics the quasiperiodic to chaotic transition in circle maps, to which I will refer and which also appears to be a reasonable metaphor for the intermittency of seismic tremor in volcanoes (Shaw and Chouet, 1989, 1991). Another type of analog system that illustrates many important regimes of nonlinear phenomena is the dripping faucet. In the literal sense, this is a device that mimics many types of simple to complex intermittency and oscillation. I'm not sure when it was first used as a model of intermittent behavior in larger and more complicated natural systems, but prior to the widely cited study by Robert Shaw (1984) the dripping faucet was proposed, in at least one geophysical context, as a device to explain the drop-like formation of plasmoids in the Earth's magnetotail (Hones,1979; Baker et al, 1990). Later I use an experimental study by Wu and Schelly (1989) in which the tuning of drop formation was sharpened by using a surfactant, revealing parallels between behaviorial regimes of dripping-faucet oscillators and other physical systems that have properties resembling those of circle maps. This includes a wide variety of phenomena displayed by volcanic oscillators at many different scales of temporal and spatial intermittency.

The above systems have a somewhat digital flavor in that they deal with discrete and discontinuous states, such as jumps between specific sites and energy levels in a semiconductor lattice, separation of discrete drops from a faucet or the magnetosphere, and motion of discrete batches of magma in volcanic intermittency. Another type of system with such digital attributes is the sandpile model of critical self-organization, such as proposed by Bak and Tang (1989) for earthquakes and other natural phenomena involving cascades of discrete instabilities about a quasi-equilibrium critical state. This is akin to the familiar angle-of-repose criterion for geological assessments of slope instabilities, such as avalanches and landslides, etc. The prototypical model of critical self-organization is the hourglass. The dynamics of the hourglass is not only visually instructive concerning critical-state intermittency, but it represents a natural connection between sandpile models, dripping faucet models, models of nonlinear clocks of circle-map affinities, and models of intermittent flow of electrons in semiconductors and magma in volcanoes, the latter giving rise to seismic (acoustic) tremor. The dynamical regimes of acoustic cavitation (e.g., Lauterborn and Holzfuss, 1991) and lasers (e.g., Harrison and Biswas, 1986) represent other phenomenologies that have affinities with these types of systems. Because these systems all involve both discrete and continuum or continuum-like states (e.g., the fluidized states of sand avalanches in the hourglass), it will not be surprising to also see affinities with the nonlinear dynamics of continua as exemplified by convecting fluids.

The dynamics of continuous incompressible fluids has provided another rich source of evidence by which to illustrate and/or explore nonlinear periodic and chaotic transport regimes. Especially informative are experiments involving forced and natural convection, including forms in which chemical or mass diffusivity plays a part in addition to diffusivities of heat and momentum. The terms "double-diffusive convection" and "thermosolutal convection" are often used for convection in which chemical bouyancy gradients are important. Examples of nonlinear behavior in convective systems are so many and diverse that no single reference does them justice. And this is without mentioning systems with chemical reactions. They, in turn, form a broad class of chemical oscillators in which the regimes of behavior are also reminiscent of those involving both discrete and continuum states. Needless to say, the periodic behaviors of such systems are

8.3

likely to be informative with regard to the possible internal dynamical regimes of multicomponent, multiphase magmatic systems.

The studies by Lorenz (1963, 1964, 1979) fall naturally within the category of convecting systems. Although they are famous for being early examples of digital computer experiments in nonlinear dynamics, Lorenz was explicit in making clear their basis in convective phenomenologies. In fact, there is a close parallel, and even identity in special cases, with the convective modes of the thermosyphon loop to which I also will refer for examples of similarities to regimes of the circle map, dripping-faucet regimes, and regimes of magmatic convection. The well-illustrated text on fluid dynamics by Tritton (1988) gives an overview of several of these topics. In the context of parallels with the circle map, experimental study of coupled convective modes by Jensen et al (1985) and Fein et al. (1985) have revealed the same type of quasiperiodic to chaotic transition as was mentioned above with regard to semiconductor experiments and the phenomena of volcanic tremor. Thus, the numerical dynamics of the circle map again provides an intermediary between the nonlinear phenomenologies of discontinuous and continuous transport systems.

A final class of relevant experiments, and one that is directly tied to historical developments in dynamics, yet oddly enough represents a sort of historical bifurcation from the thermomechanical effects I want to emphasize, is that of the simple mechanical oscillator. In this category are the phenomena of the nonlinear pendulum (in many forms), the phenomena of coupled springs and masses with forcing and damping, and phenomena of vibrating cantilever beams and filaments, flutter of aerofoils, and so on. Again nonlinear dynamics makes evident conceptual parallels with the other types of experiments mentioned. A conspicuous one is the formal resemblance of the nonlinear pendulum to the convective thermosyphon loop, hence also providing a rather direct connection to circle maps and to the Lorenz system, especially when it is described in the form of a waterwheel (e.g., Lorenz, 1979; Sparrow, 1982, Appendix B).

Nonlinear dynamics has revealed what can be seen in retrospect as ironies or paradoxes in the development of physical theory. Some may be evident in this introduction. It took the use of the digital computer as an experimental device for dynamical exploration to dramatize numerically the richnesses of periodic and chaotic regimes, their self-similarities and universalities. Furthermore, among the varieties of especially dramatic experimental demonstrations now known are the phenomenologies of electronic transport in semiconductors in which the nature of the energetic states is intrinsically involved with lattice dynamics and thermally activated degrees of freedom. Nearly as complex in detail, and more so in terms of an inability to constrain and/or measure all of the mechanical degrees of freedom, not to mention the thermal ones, are the experiments on convection and on chemical oscillators. By contrast, the classical studies associated with pioneering turn-of-the-century work in nonlinear dynamics by Poincaré, Liapunov (several different spellings exist among widely cited sources), Duffing, and Birkhoff came from notions associated with the mechanical and/or mathematical theory of oscillations (see Abraham and Shaw, VISMATH 1-4, 1982-1988).

It seems possible that experimental demonstrations of nonlinear attractors and chaotic dynamics might have emerged sooner had ideas of coupling between thermal and mechanical fields become more evident to physicists during the time following Rumford's and later Joule's experiments (fascinating historical sketches of their experiments, and the reactions of contemporaries such as Kelvin, are given by Sandfort, 1962). Brief references to these experiments in college physics textbooks do not do justice to the implications of their work from the standpoint of nonlinear phenomena. Had they done so, I for one would sooner have become aware that in the study of geophysical transport processes, and especially magmatism, I was intrinsically involved with nonlinear dynamics in the forms that have been revealed only during the last two decades. In this and other respects I have some empathy with the epistemological remarks by May (1976). Although I was aware of nonlinear behavior in early work on the rheology of magma (e.g., Shaw, 1965, 1969; Shaw et al., 1968), I was not sufficiently conversant with mechanical theories of oscillation to make that connection graphic. A concept of explicitly coupled nonlinear oscillation was invoked in the hypothesis of tidal generation of magma, with I. J. Gruntfest (Shaw, 1970; Gruntfest and Shaw, 1974), but its interdiciplinary implications only became vivid later following the mentioned work of Lorenz, and that of May (1976), and Feigenbaum (1980).

Some aspects of nonlinear dynamics applied to magmatic phenomena have been reviewed previously (Shaw, 1987a, 1988) and have also been invoked as an essential basis for interpretations of periodicities in the geological record (Shaw, 1987b). In those reviews, however, I did not adequately bring out the commonalities among the different types of systems that may already be evident in this introduction and that the rest of this chapter addresses. The leitmotif of coupled oscillation, exemplified by the mathematical metaphor of the circle map as prototypical coupled oscillator or nonlinear clock mechanism (I would include notions of 'biological clocks' in the same vein) was touched on but was not fully appreciated until it became evident as the fundamental mechanism in studies of volcanic tremor with B. Chouet (Shaw and Chouet, 1988, 1989, 1991; Chouet and Shaw 1991).

8.2 Thermal Feedback, Chaotic Oscillations, and Intermittency

8.2.1 Magmatic processes and Rumford's Canon.

Unfortunately, space does not permit a description of the experiments performed in 1798 by Benjamin Thompson, an American whose Tory sympathies led him to follow a career in England and eventually to supervise the operation of a cannon-boring mill in Bavaria, where for various services he was given the title of Count Rumford. Apparently he was at heart an experimentalist, because he was intrigued by the intense heating of the brass cannon and its cuttings by the boring tools, especially when they became dull! He came close to measuring the mechanical equivalent of heat in an experiment where he built a vat filled with water surrounding a vertically-mounted cannon barrel and measured the time required for a team of horses to turn the borer through enough cycles to raise the water temperature to boiling. Had he used pulleys and weights to calibrate the rate of work done by the horses, he would have essentially duplicated the later experiment by Joule in which he used a paddle wheel driven by pulleys and falling weights to vigorously stir a container of water in which he carefully monitored the temperature (e.g., Sandfort, 1962, p. 78). Ironically, it was also by such a pulley-driven device that my coworkers and I were able to measure the nonlinear viscosity of molten basalt in a ponded batch of lava in Hawaii in 1965 (Shaw et al., 1968). And it is equally ironic that it was just such an experiment as that performed by Rumford by which I first became impressed by the nonlinearity of thermomechanical coupling, further influenced by a collaboration with I. J. Gruntfest who had pioneered the concept of thermal feedback in various rheological contexts (see Gruntfest and Shaw, 1974), as well as by other experiments in the laboratory reported in Shaw (1969).

Even though thermodynamics eventually evolved to become a dominating influence in the physical sciences, a curious bifurcation has persisted in terms of the separation of mechanical and thermal parameters in dynamics. By speculation, it is as though the influence of the old "caloric theory", according to which heat represented a substance separable from that of the mechanical device with which it was correlated, has continued to haunt concepts of viscosity and its relationships to dissipative motions; for example, see the descriptions in Sandfort (1962) of the reluctance with which that theory was abandoned even after the work of Rumford, Joule, and others between, had demonstrated the intrinsic nature of heat as a manifestation of chaotic motions produced in a fluid by the rate of doing mechanical work. The principle of energy conservation became established, but few after Rumford seemed curious about what was going on at the locus of heating, for instance at the interface between borer and cannon barrel and in the fragments produced by the boring process, not to mention the trajectories of *their* motions as functions of boring rate, tool sharpness, and so on. Everyone from primitive times on has been aware of what happens when you spin a stick against a block of wood, slip a rope through your hands, or repeatedly flex a piece of wire, but few dynamicists have asked how this same phenomenon will affect the evolution of *local* trajectories in a mechanical experiment in which the principles of thermodynamics require that some form of thermomechanical dissipation must be heterogeneously influencing the loci of interaction. This effect was dramatized in my own experiments in 1965 - 1969 because of the proximity of the working substance to its range of multicomponent, multiphase melting temperatures, and because the rate of motion (relative rotation rate of coaxial cylinders in a viscometer) was extremely sensitive to the change in viscosity produced by the combined effect of changes in temperature and changes in crystal-liquid proportions (cf., Shaw, 1969).

In 1967, in order to visualize these thermal effects, I performed an experiment that was directly analogous to those of Rumford and Joule. However, I substituted a fluid that had a viscosity more like that of molten lava (e.g., resembling that of slow-moving pahoehoe, or "ropy", lava shown in movie footage of typical Hawaiian eruptions). The results produced the expected heating effects, but they also demonstrated the nature of feedback coupling between the local rate of shearing and the local rate of heat transfer, in this case by a combination of conduction and convection combined with some endothermic cavitation (formation of vapor bubbles near the locus of shearing). Some results of this experiment are described for the first time below, because they fit with the theme of dissipation combined with the simultaneity of differing types of dynamical model evidenced in the same system and even in the same general process. In this context, the notion of coupled oscillation on many scales of time and size is unavoidable, because at the one extreme there is the excitation of strong thermal oscillation at molecular to macroscopic scales, and at the other extreme there is a variable and complex system of oscillations of the entire fluid as shown by trajectories of neutrally bouyant particles within the artificial boundaries established by the laboratory scale of the system, a space occupying roughly 10 liters of volume (10^4 cm^3). Removing the boundary constraint, and considering the cases of magma in large chambers and magma at source depths of melting in the planetary interior, it can perhaps be seen how much these experiments influenced my views on the effects of viscous dissipation on modes of deformation and magma transfer in planetary bodies, as reported in Shaw (1969, 1970) and Gruntfest and Shaw (1974). From the standpoint of energy conservation and net thermal history, similar arguments have been applied to the satellites of Jupiter (e.g., Cassen et al., 1982).

8.2.2 Nonlinear dynamics of viscous dissipation in a silicone fluid.

I will summarize these experiments with illustrations shown in Figures 8.1 through 8.9. Figure 8.1 sets the scene for those that follow. Here a cubical vat made of clear plastic

8.6

measuring about 25 cm on a side was filled to 16 cm depth with a silicone oil that had a nominal viscosity of 1000 poise (100 Pa·s). A cylindrical rotor of varying diameters, driven by a constant speed motor, could be lowered into the fluid to variable depths, much like a drill press. The motor was calibrated to read torque as an output voltage on a linear-log strip chart that gave a continuous record over three decades of torque during manually adjusted changes in speed. Temperatures in the silicone fluid were measured with small-diameter thermocouples at various locations and at the transparent walls of the vessel. The walls remained nearly isothermal at room temperature during the experiments, and the top and bottom of the vessel were relatively insulated compared to heat loss through the sidewalls. Hence, the greatest rate of heating was at the center, and cooling was dominantly at the vertical walls. In Figure 8.1, in order to crudely follow temperature changes at the rotor interface, a thin-walled mercury thermometer of about 0.5 cm outside

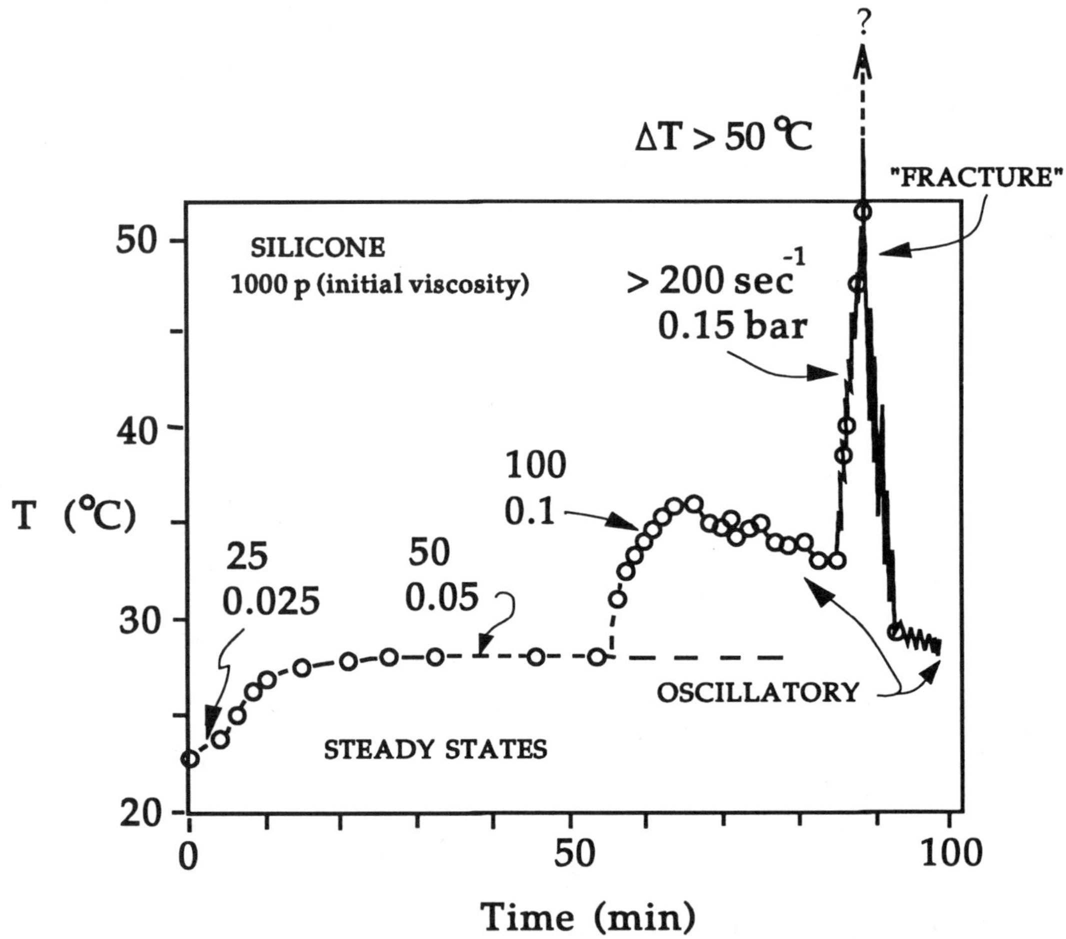

Figure 8.1: Experimental rotational shear of a silicone fluid in which the dissipation of mechanical work is converted to heat, inducing thermomechanically coupled cavitation and liquid fracture. Experiment performed by the author.

diameter was used as the rotor itself. It was lowered to within about 3 cm of the bottom. There was no evidence of slip at its surface, because the temperature readings of the thermometer were systematic and the locus of most intense cavitation and eventual shear failure were situated within the fluid at about 2 to 5 cm from the thermometer surface. The

8.7

speed, and torque, were advanced in stages with the results shown. When oscillations in temperature and torque began, the speed was adjusted to maintain an approximately constant torque. At a shear stress at the rotor of about 0.15 bar (0.015 MPa), the temperature rise abruptly accelerated, and the speed could not be advanced fast enough to maintain constant torque. Simultaneously, the tenuous cloud of cavitation that had become more and more conspicuous above about 0.05 bar (still, however, amounting to only a few percent bubbles) grew rapidly over a cylinder of roughly 10 cm diameter about the rotor. Within that cloud bubble rings and arcuate cylindrical sections of mirror-like rupture surfaces began to appear, eventually with an abruptness too fast to time (<< 1 sec). Almost as suddenly the arcuate surfaces of rupture disappeared as the torque fell and motor speed went to its maximum (about 3000 rpm).

The temperature transient of the thermometer exceeded 30 deg C, and the actual temperature rise at its surface must have been much greater. The initial stress-drop was 70 to 80 %. The speed was maintained for an hour or so while the output voltage was recorded, giving the history of intermittent stress oscillations shown in Figure 8.2. For reference, oscillations are shown at a lower speed and stress level, corresponding to a central temperature rise of < 5 deg C. Even at that rate of shear there was some local nucleation of cavitation bubbles associated with stress oscillations of roughly 1 sec average period (similar to the dominant frequency of volcanic tremor and reasonably interpreted as acoustic cavitation). Liquid "fracture", though novel, is known in other contexts and is the essential mechanism of the cavitation process. The interesting feature here was the implication that the rupture surface grew within the cavitation cloud by a cascade of failures between bubbles, a mechanism not unlike some models of earthquake nucleation, and reminiscent of the sandpile model of self-organized failure.

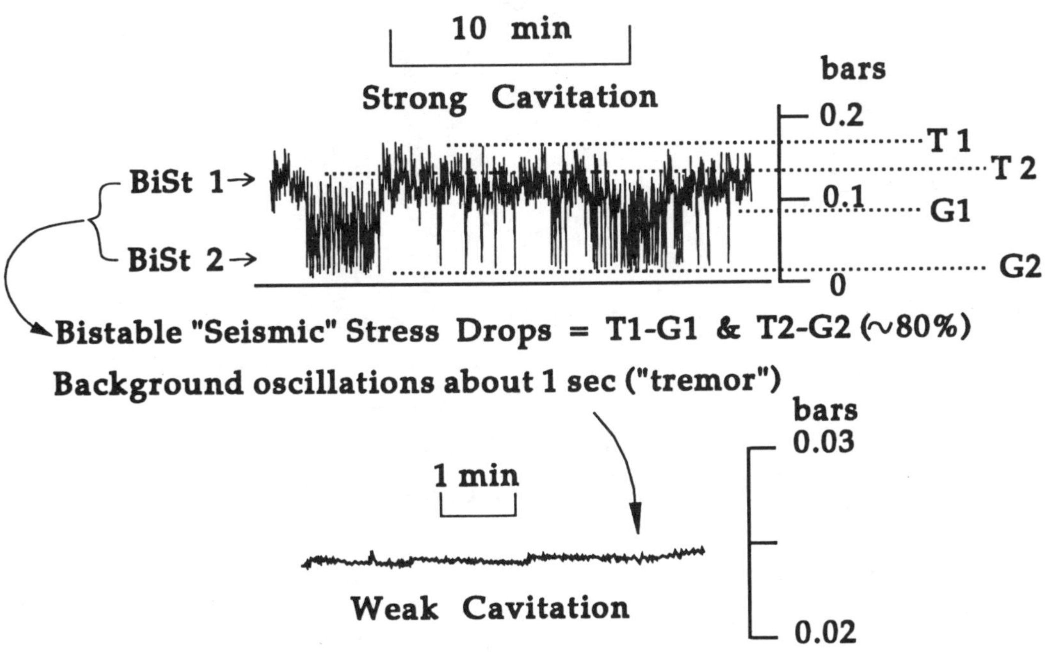

Figure 8.2: Time series of cavitation events in Fig. 8.1 (see text).

Two particular aspects of Figures 8.1 and 8.2 are relevant to models of intermittent and chaotic attractors discussed later. One is the quasisteady oscillatory regime in Figure 8.1 at about 0.1 bar shear stress (roughly a 10 deg C temperature rise at the center with

8.8

fluctuations of a few degrees), which relates to variations in toroidal flow velocities resembling those of the oscillatory regime in the simple convection loop and thermosolutal convection. The other is the roughly bistable pattern of oscillatory intermittency in Figure 8.2 which resembles the types of time series seen in other bistable systems, such as mechanical oscillators and the convection loop.

A schematic view of the convective pattern in vertical section is shown in Figure 8.3. The 'flow lines' are the loci of a neutrally bouyant colored marker about 1mm in diameter read from half of one face of the container against an orthogonal grid placed behind it.

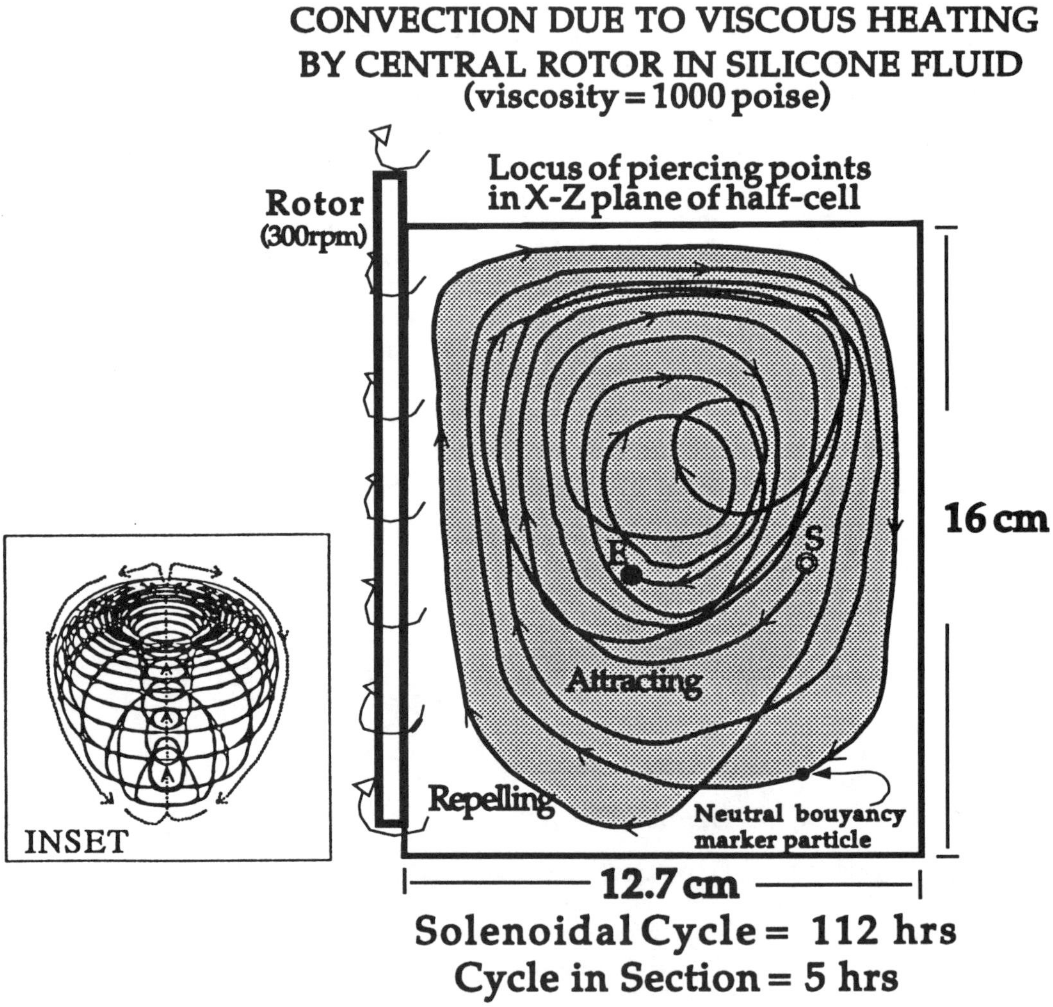

Figure 8.3: Trace of a neutrally bouyant marker in convection driven by dissipative heating, as read in a vertical section centered in right half of cell. The total power generated at this setting was roughly 4 watts, corresponding to the vicinity of the steady and oscillatory regimes of Fig. 8.1. The inset is a caricature of the overall circulation. See text for discussion.

Thus, the lines are the locus of piercing points through an invisible plane placed vertically at the center of one side of the tank. Because of the viscosity and rotational geometry, the motion of the marker varied greatly in velocity, taking long times to make one revolution about the rotor when near the side while it rotated rapidly when near the center (but never closer than 2 or 3 cm from the rotor). Although I attempted to monitor every revolution, clearly this was not feasible over the nearly 5 days of the readings. Therefore, Figure 8.3 is an odd sort of *Poincaré section* (strictly, the Poincaré section represents points on an attractor read at the period of forcing, or at some constant delay time, like a stroboscopic recorder). Implicitly, the number of piercing points are dense near the center and tenuous near the sides, yet the lines nearer the sides are more representative of the actual migration per cycle. It may be noted that any finite set of readings would be a *fractal dust*, concentrically filling the shaded area of the attractor from inside out, but only after infinite rotations. Because the trajectories are intertwined the distribution actually would be clustered, hence *multifractal* as discussed by Shaw and Chouet (1988, 1989) relative to volcanic tremor, coupled semiconductor oscillators, and coupled convective oscillators.

Clearly, the motion of the marker in this experiment traces out a geometric object of great complexity. The inset shows a schematic of the general circulation (into the center at the bottom and at different heights, and out at the top and at different heights, forming a waterspout-like motion but in which trajectories also tend to fill the body of the object with a tangle of crossings that are extremely dense at the center and tenuous at the margin). If rotational velocity vectors are ignored, then the trajectories would look like a *chaotic toroidal circulation* (intertwining of trajectories within a donut-shaped space). Ignoring the toroidal motion and considering only the trajectories of rotation, the object would look like a chaotic coil or *chaotic solenoid* in which the windings are intricately intertwined, of course being most dense at the center. There is no doubt that the object is an attractor, where the grey area is the *attractor basin*, because the marker was *repelled* both from the vicinity of the rotor and the walls. The topological complexity and the intertwining also point to a chaotic motion even though the flow regime as a whole is nominally "laminar", a term that perhaps is not very useful in magmatic fluids because there can be very intricate chaotic tortuosity in highly viscous subcritical flows (low Reynolds-number 'creeping flows' in the parlance of fluid dynamics).

Thus, Figure 8.3 represents both a *chaotic attractor*, as well as a very *strange attractor* indeed. The term chaotic just means that the trajectories are not demonstrably periodic (i.e., they do not repeat by reoccupying exactly the same path after arbitrary numbers of cycles; obviously this is impossible to prove in any finite experiment, a point easily illustrated with the circle map where one can tune to *mode-locked rational periods* and *irrational quasiperiods* of such intricacy that it is impossible by inspection or even power spectral analysis to demonstrate that the trajectories repeat, within the precision of calculation; cf. Shaw, 1987b). The form of the meanderings indicates a very low likelihood that any one section of path in Figure 8.3 would ever be duplicated exactly. The term "strange" was originally applied to chaotic attractors in which the trajectories and/or Poincaré sections demonstrably formed a fractal distribution, usually taken to be an invariant fractal but now recognized to be multifractal in complex attractors associated with actual flows (cf., Jensen et al., 1985; Sreenivasan and Meneveau, 1986).

The nature of the circulation is clarified in Figures 8.4 and 8.5 which are analogous to Figure 8.3 but are projections to a horizontal section, as though looking down the axis of rotation. Figure 8.4 schematically shows the way in which the marker spirals in and out as it both moves up and down and 'spins' about that axis. The initial part of the second cycle is shown to indicate how the path of the marker will diverge from its previous path. If two particles were originally nearby, they would increasingly differ in location and path, a mode of behavior called *sensitive dependence on initial conditions* first made famous in Lorenz's (1963) study and sometimes called the "butterfly effect" (I'm not sure if that refers to the form of his strange attractor which looks a bit like a butterfly, as shown later in Fig. 8.24, or to someone's apocryphal remark that 'even the flapping of a butterfly's wings will change the outcome of a forecast in the chaotic regime'). Although I was not aware of Lorenz's work at that time, I tested the idea of divergence by inserting a second marker of different color and noted that it did not duplicate the path of the original one even though their projected paths crossed a number of times and occasionally even approximated the same segment of path (separated in time) over short distances.

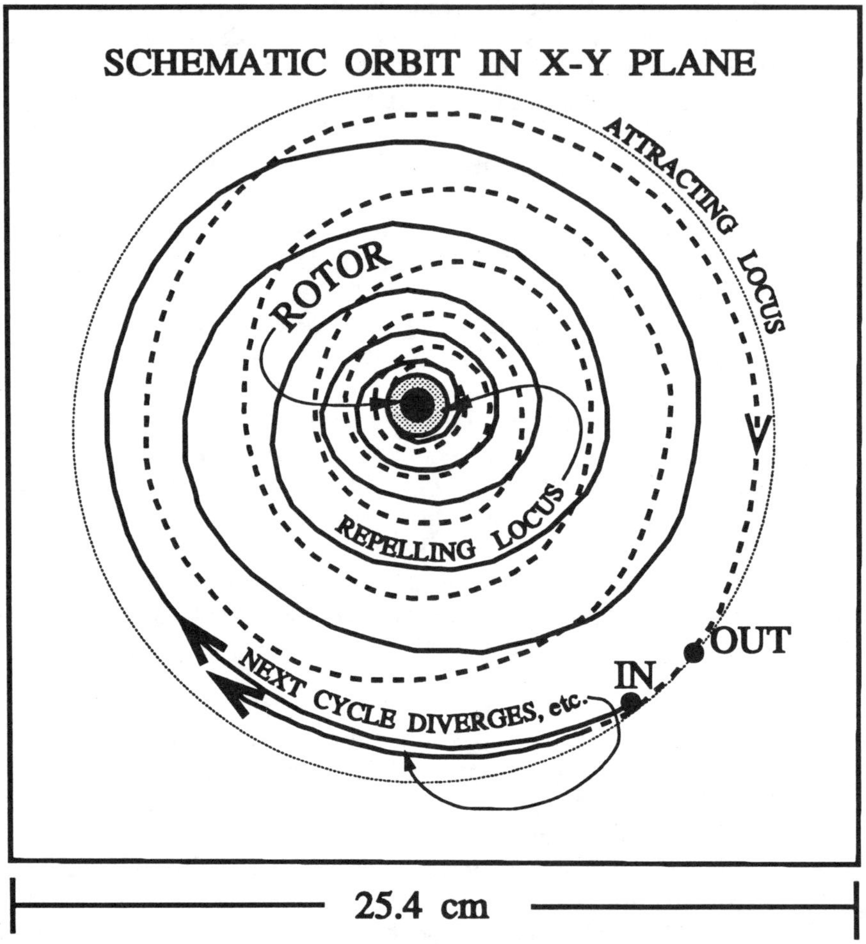

Figure 8.4: Schematic view of the general circulation in the silicone fluid looking down the axis of rotation. The circle marked "attracting locus" is the general limit beyond which the marker did not travel toward the walls of the container. Similarly, the "repelling locus" near the center demarks a zone of exclusion near the rotor (black circle). See text for discussion.

An even more revealing view is shown in Figure 8.5, which is analogous to the pseudo-Poincaré section of Figure 8.3. The dots are the locations of one of the markers at different times as it 'orbited' the rotor (the 'primary'). Thus, the paths are analogous to the motions of a satellite or of a 'planet' in a very chaotic 'solar system', where the 'yearly' apogee and perigee vary in the extreme and gyrate in succeeding cycles (cf., Wisdom, 1987). The progression of these distorted elliptical orbits also resembles the motions of the elliptical paths of a nonlinear spherical pendulum projected to its base, as will be shown (see Miles,

8.11

1984; Tritton, 1988, p. 402). This visual analogy is reinforced by the fact that the observed rotation speed ('orbital velocity') of the marker about the center varies with the radial distance somewhat like Kepler's law out to about the limit of the attracting region of Figure 8.3 (a bit less than 1 cm from the wall), after which it would fall off rapidly due to the wall effect. The rotation rate vs distance was determined by timing the angular motion of markers over short arc lengths using a stopwatch.

ORBITAL WANDERING OF PARTICLE
(Looking down axis of revolution about rotor; 3.2 revolutions about 'primary' during 112 hrs, or about 36 hrs in average cycle, or 'year')

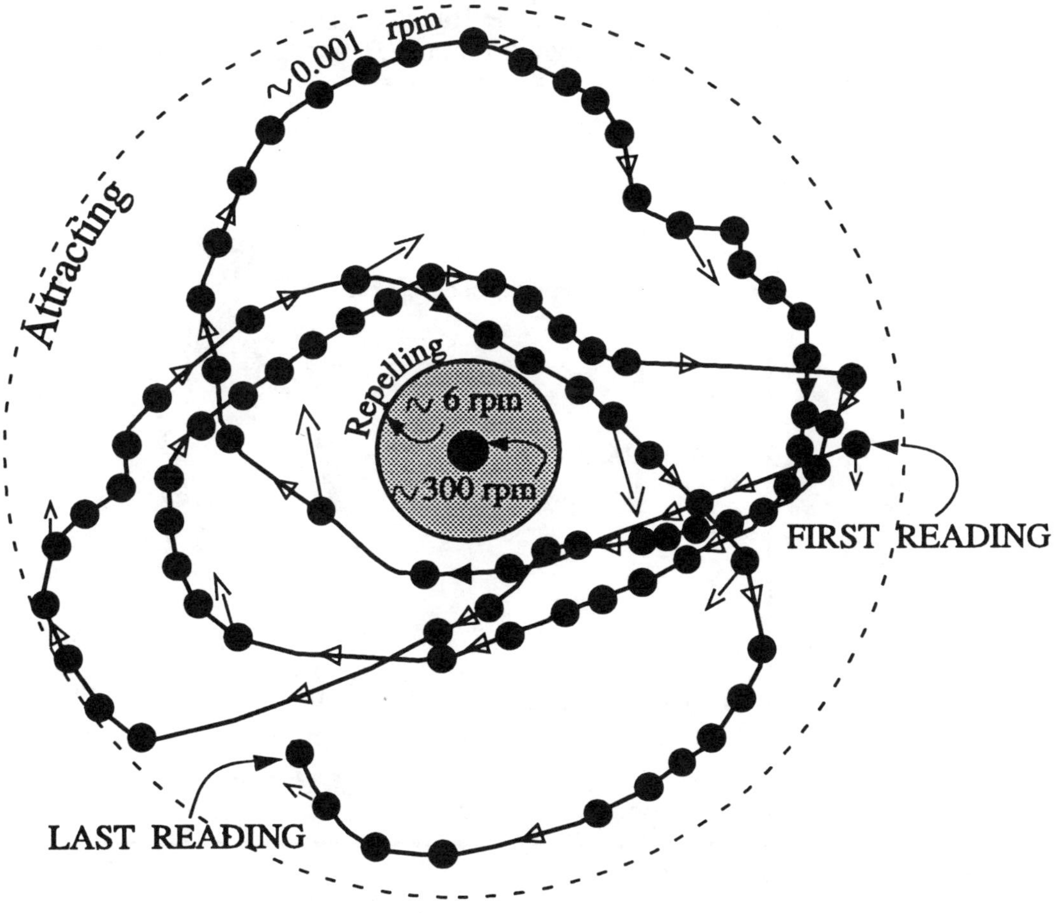

Figure 8.5: Mapping of the radial position of the neutrally bouyant particle during the course of the experiment (112 hours). Readings could not be taken at equal intervals of time because of the greatly differing rates of motion near the center and the outer limit of the attracting locus, as identified in the figure by schematic vectors and rotation speeds (see text).

Incidently, calculations of chaotic orbits of asteroids over millions of years have been carried out by Wisdom (1987) expressed as variations of eccentricity, the measure of deviation of conical sections (ellipses) from a circular orbit. Some of his results bear an even stronger resemblance to the properties shown in Figure 8.5, and a regime of bistability exists which bears a marked resemblance to the time series in Figure 8.2 (see Figs 8.41 and 8.42). The latter would correspond to a *bifurcation* in Figure 8.5 in which circulations near the center would intermittently alternate with paths like those shown. This could in fact exist in that figure if the 'strobe' points had been plotted at much smaller time intervals and might be searched for in the time record, but so far I have not made the attempt.

The traditional phase portrait represents trajectories obtained by plotting a position coordinate against its time derivative, and the Poincaré section is its stroboscopic form. Figure 8.6 shows a Poincaré-like variant of such a section in which radial velocity is plotted against vertical position, revealing a fractal-like pattern of higher positive velocities (outward) at the top of the attractor relative to negative (inward) velocities near the bottom (there are insufficient points to determine the fractal dimension). The time series of both the radial and vertical positions are shown in Figure 8.7, and should be compared with time series of other chaotic convective motions shown later.

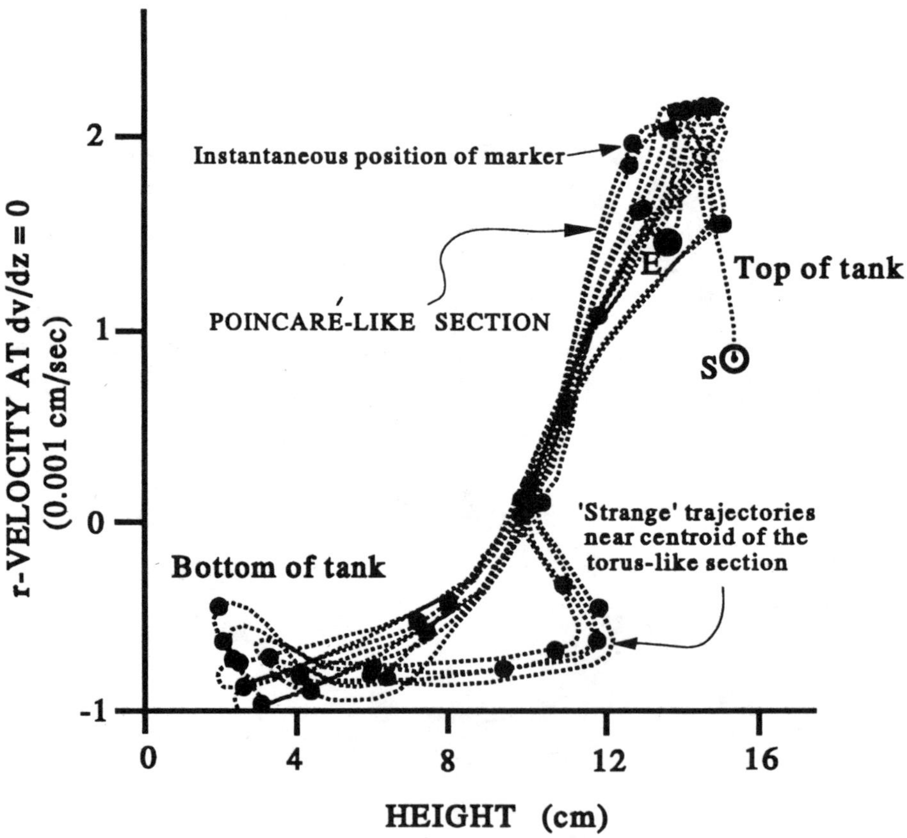

Figure 8.6: Poincaré-like section of the radial velocity of the marker versus vertical position. Compare with the projection shown in Fig. 8.3, and discussion in text.

To conclude this description, the next two diagrams plot the data in a form more appropriate to conventional portrayals of convection regimes. In Figure 8.8, temperature is plotted against the speed of the rotor, the latter representing the total power driving the

8.13

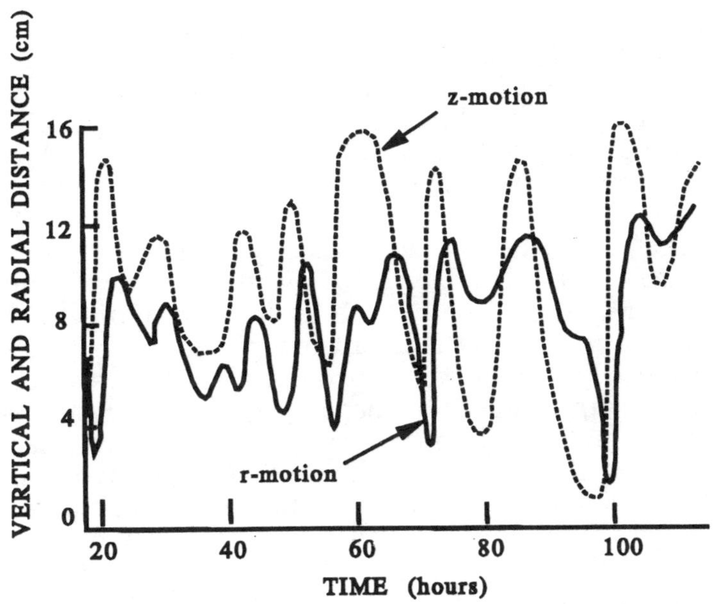

Figure 8.7: Time series of the radial and vertical motions of the neutrally bouyant particle in the experiment of Fig. 8.3.

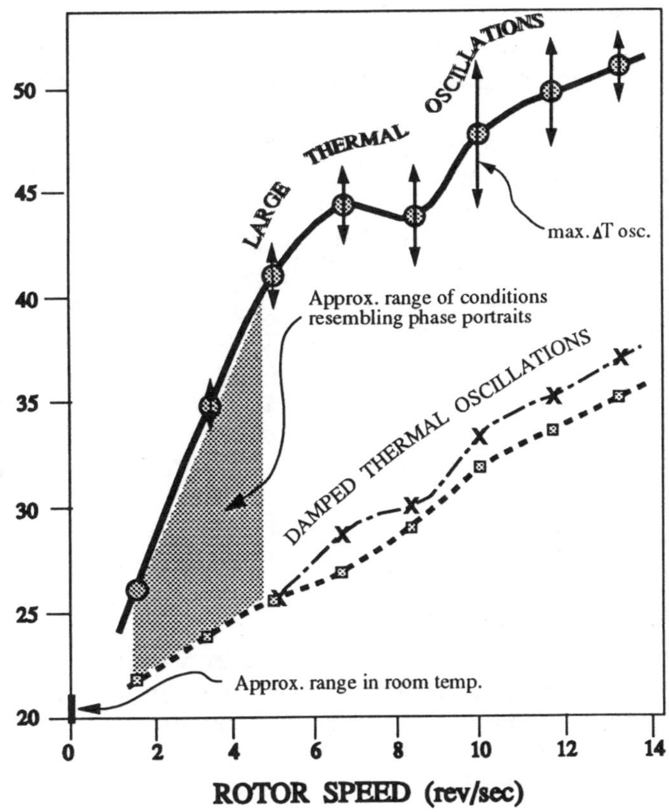

Figure 8.8: Temperature oscillations in the fluid as a function of rotation speed, at the center (circles with arrows), at the convective null point (crosses), and near the walls (squares).

convection. Thermally, the mechanical input is analogous to placing a cylindrical heater at the center of the tank (this was also done to compare with the power of the mechanical source, giving good agreement for heat transfer parameters). Thermal oscillations are shown by double-headed arrows. The conditions of Figures 8.3 to 8.7 represent a chaotic mode near the beginning of the oscillatory regimes (more complicated forms of bifurcations are likely beyond the grey area, as will be seen in other portraits of chaotic convection).

Figure 8.9 describes convection in the usual way with a plot of nondimensional heat transfer, given by the Nusselt Number (N_T), against the nondimensional measure of thermal forcing, the Rayleigh Number (R_T). N_T essentially measures the ratio of convective transport, governed by the thermal input power, to conductive transport, in turn governed in the quasisteady state by the temperature gradient across the container. R_T is a measure of bouyant forcing produced by the temperature rise at the center relative to the viscous resistance to the motion. The absolute values in the figure, relative to other plots of this kind, are not necessarily significant because they depend on rather poorly resolved length ratios in cylindrical systems. For instance, the analysis of convective heat transfer in long vertical tubes by Lighthill (1953), applied to the problem of cooling turbine blades, appears to agree fairly well with these results, but the aspect ratio is not precisely defined in the rotational geometry. Nevertheless, the oscillatory 'bucket' he identified as likely corresponds to the present oscillatory regime rather well, considering the different nature of thermal forcing in the two cases. I will return to this in the oscillatory regime for the case of thermosolutal convection where calculations by Knobloch et al. (1986), analogous to those of the Lorenz system, reveal details of multiple periodic and chaotic bifurcations in flow regimes. These, in turn, bear resemblances to bifurcational regimes of the experimentally tuned dripping faucet system of Wu and Schelly (1989).

I have given this much space to the above experiment because it stands as a metaphor for magmatic processes and chaotic attractors of natural systems. Of paramount importance in my view is the fact that the various types of thermal and thermomechanical instabilities are introduced naturally by the deformation field itself rather than being imposed as boundary conditions. Even though the net results are similar in terms of heat transfer budgets, the occurrences of other instabilities that accompany that transfer are directly related to the feedback between viscous dissipation, mass transfer, and criteria of material instabilities (such as cavitation and fracture). In a sense, this experiment is 'inside-out' compared with the conventional approach to convective instabilities. Here, thermal excitations are consequences rather than impositions. The inverse would not in itself predict the effects of cavitation and fracture, and probably would not predict the same type of chaotic intermittency for the same net thermal power.

The analogous convection cell driven by a cylindrical heater will produce analogous oscillatory convection, but otherwise it does not have the same universal implications. The above experiment could alternatively be described as a system of coupled oscillators, a point that is central to concepts of chaotic intermittency in discontinuous systems such as the dripping faucet and sandpile models, as well as seismic tremor, acoustic cavitation, volcanic intermittency, and so on. For example, depending on how one views the natural periods of the system, the Rumford-like modes of excitation of the silicone fluid possess several interacting frequencies: (1) There is the spectrum of rotational frequencies as a function of radius; (2) The toroidal circulation in the plane of Figure 8.3 represents one of the characteristic natural frequencies; (3) There is a solenoidal frequency associated with approximate returns through the windings of the combined toroidal-rotational path; (4) The gyration of orbital parameters of Figure 8.5 also has a characteristic, and longer, average return time; (5) A number of different frequencies are associated with the modes of cavitation and 'seismic' failures of Figure 8.2; and (6) Bouyancy pulsations related to waxing and waning cavitation clouds (not explictly discussed) add other frequencies.

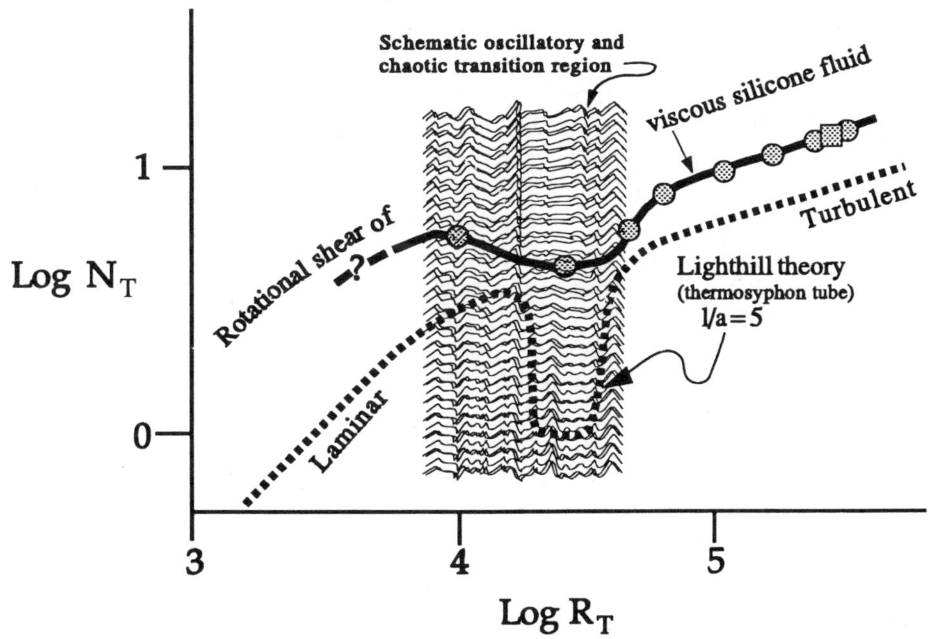

Figure 8.9: Plot of heat transfer expressed in the conventional way as the Nusselt number plotted against the Rayleigh number. The wavy pattern indicates the region of relatively steady oscillatory convection, compared with the theory of Lighthill (1953) for vertical cylinders. The square gives the result for substitution of an electrical heater with the same equivalent power as viscous heating.

Although these are all driven by the forcing frequency of rotation, their interactions are analogous to natural systems in which it is often not possible to distinguish between "forcing frequencies" and "natural frequencies". In my view, this makes the experiment a powerful analog for chaotic motions in planetary bodies, especially when the imposed mechanical source of rotation in the experiment is associated with the total mechanical work of deformation done on a planetary or celestial object. Although radiative thermal inputs sometimes allow us to neglect mechanical ones, and in other cases mechanical inputs overshadow thermal ones (e.g., as in classical mechanics), it is the thermomechanical coupling of inputs that appears to be essential to the character and details of the nonlinear periodic and chaotic attractors that are inevitably generated in complex systems. Despite the complexities, the strong resemblance of net convective modes to simpler systems also argues for an important role of self-organization and universality, aspects that we can watch for in other systems examined.

A most important demonstration of these experiments, in my view, concerns distinctions between deformations at constant stress vis a vis constant speed of displacement (rotation rate in this case). Neither idealization persists in nature, unless it is an outgrowth of self-organization averaged over some characteristic cycle time of stress relaxations and/or shear-rate relaxations (intermitttent instabilities). To be observed, in either case, their scaling must be commensurate with that of the physical measurements. In both cases, the local shear rate is not constant because it adjusts to the local thermal activations as well as to the usual mechanical effects. This interaction is coupled directly with criteria of irreversible thermodynamics at the molecular level (local equilibria of chemical reactions) and is the principle reason for permitting thermomechanical coupling to be present in an experiment, a theoretical model, or in nonlinear calculations. Some workers have discounted thermal feedback instabilities in nature because they mistakenly assume that they are significant only at constant stress, and that this condition is not realistic. Other workers have argued for a constant speed (or creep rate) boundary condition, thinking that this will stabilize the system by negative feedback and avoid instability cascades (runaways). Figures 8.1 and 8.2 belie such assumptions. When the system reached constant speed at the end of the run in Figure 8.1, the behavior was not stabilized, unless one takes an average over many cycles of intermittency in Figure 8.2. In that figure the rate of change of stress becomes extreme or discontinuous at each of the 'seismic' events at both the smaller and larger scales, expressing the earthquake-like fracture phenomena described. Such effects are consistent, for example, with an almost ubiquitous presence of fracture in otherwise homogeneous cosmological objects (their relative fluidity only affects the time scales of the instabilities). Satellites, such as our Moon and the satellites of Jupiter, which otherwise might not generate gravitational stresses exceeding static fracture criteria, might sustain instabilities of these types (e.g., Gruntfest and Shaw, 1974; Wones and Shaw, 1975).

The off-scale peak instability of Figure 8.1 is of generic type for systems that are overdriven relative to rates of endothermic changes of thermodynamic phase states. They are sometimes called "thermal runaways", or *thermomechanical catastrophes*, and can result in solid-state phase changes, melting, vaporization (cavitation, vesiculation), and fracture, or some combination of such instabilities. Melting and vesiculation, or direct degassing of solid-state assemblages at mantle depths, may be typical reactions in terrestrial planets and analogous satellites of the outer planets. Volcanic processes offer many opportunities for such effects both in near-surface phenomena associated with intrusion and eruption and in melting phenomena at mantle depths of primary magma production. Diatremes and kimberlitic pipes ('diamond pipes') are phenomena where the dynamics of the solid-gas phase change predominates, with or without the participation of the melting reaction. It seems likely that analogous processes would operate in plasma systems. Solar flares and stellar jets have some striking resemblances to eruptive instabilites. The former includes tremor-like fluctuations with frequencies and fractal attractor dimensions resembling those of near-surface volcanic tremor (cf., Kurths and Herzel, 1987; Chouet and Shaw, 1991).

8.3 Periodic and Chaotic Regimes of Simple Nonlinear Models.

WARNING: NONE OF THE ILLUSTRATIONS THAT PORTRAY RESULTS OF OTHER STUDIES ARE QUANTITATIVELY ACCURATE. THEY HAVE BEEN SCHEMATICALLY REDRAWN TO EMPHASIZE ONLY THEIR GENERAL FORM. ORIGINAL SOURCES MUST BE CONSULTED.

8.3.1 Prefatory Note On Complexity in Simplicity.

The following sequence of illustrations recapitulates some properties of several well-known nonlinear mechanical models with simple geometries. Perhaps the greatest irony and paradox of nonlinear dynamics is that even the simplest of numerical or mechanical systems is capable of effectively unlimited complexity, witness the pioneering discussions by Lorenz (1964), May (1976), and Feigenbaum (1980), each of which foretold in their own way the coming of a major scientific revolution, still growing, especially with regard to concepts of complexity, statistics, and universality. Minds are boggled at the notion that systems characterized by totally different constitutive relationships and of totally different scaling magnitudes in time and space can have not only analogous but essentially homologous structural identities. Yet we all have accepted this kind of universality with regard to relationships between plane and spherical geometry, at least until Mandelbrot's concept of *fractal geometry* was recognized as being part and parcel of the paradigm shift with the advent of *fractal attractors*. Period-doubling sets and other periodic sets, such as mode-locked and quasiperiodic sets of the circle map are fractal, as well as the "strange" attractor coined by Ruelle and Takens (1971, p. 170) because it was "locally the product of a Cantor set and a piece of a two-dimensional manifold". The Cantor set is one of the prototypical cases of a 'dissection set' or 'disconnected fractal'. Although Ruelle and Takens did not use the word fractal, the context was clear; Mandelbrot (1982) acknowledges that the word "fractal" was coined in his French edition of 1975, but he had been writing about "fractional dimensions" at least since 1967 when he published on the problem of the length of coastlines, following L. F. Richardson's studies of the problems of the lengths of irregular political boundaries and social conflicts (his 'strange book' called *Statistics of Deadly Quarrels*, cited in Mandelbrot, might well be considered as a part of everyone's highschool education, especially would-be psychologists and politicians; see Abraham, 1990, where psychological phenomena are discussed in the light of nonlinear dynamics).

Recognition of the potential complexities of simple systems raises a dilemma: if simple models can be arbitrarily complex, what is the point of studying complex systems? In a statistical sense the issue is sometimes moot (cf., Lorenz, 1964). To me a central point appears to be that structural regimes of periodicity of different types and chaos are subject to subtle tuning effects, to coupling phenomena, and to universal principles of self-organization. Hence the fabric by which the properties of simple systems are woven together in phenomenologically richer systems is a problem of great taxonomic interest as well as one involving theories of systems networks, neural network models (e.g., Kepler et al., 1990), cellular automata, and so on. It is therefore of interest to examine some of the similarities between regimes of behavior seen in composite systems of the sort described as a step toward seeing other analogies and homologies with broader classes of natural phenomena.

8.3.2 The nonlinear pendulum.

To begin with, we can look at the mechanical pendulum and ask what might happen if it were to encounter effects of thermal feedback. Figure 8.10 shows the linear mode with oscillations in the vertical plane, while Figure 8.11 represents a modification that permits additional forcing and the possibility of internal friction. A bridge is added that introduces flexural strain of the suspending wire above a critical amplitude of swing. Because the pendulum is in an evacuated container, forced flexing of the wire will cause heating at the hinge point. Heating will have two effects, one being a lengthening of the wire with a consequent change in the period of the motion. Also, the dissipation causes additional damping to that already present. At constant forcing, therefore, the change in period caused by these effects, combined with the effect of the shortened arm length beyond the critical amplitude, will eventually decay as heat gradually leaks from the system and it returns to its original mode. With continued forcing, the pendulum may then oscillate bistably between the relatively linear and nonlinear modes (cf., Wisdom, 1987).

Any departure from the vertical plane in the nonlinear mode, however, will also induce a component of spherical or conical motion. This can also be induced in the linear mode by introducing a periodic translation of the point of suspension. If the translational oscillation is near the natural period, chaotic motions in the spherical pendulum domains occur without the added effects of frictional dissipation, as studied by Miles (1984). His work documents a rich spectrum of periodic and chaotic orbits of the nonlinear spherical pendulum.

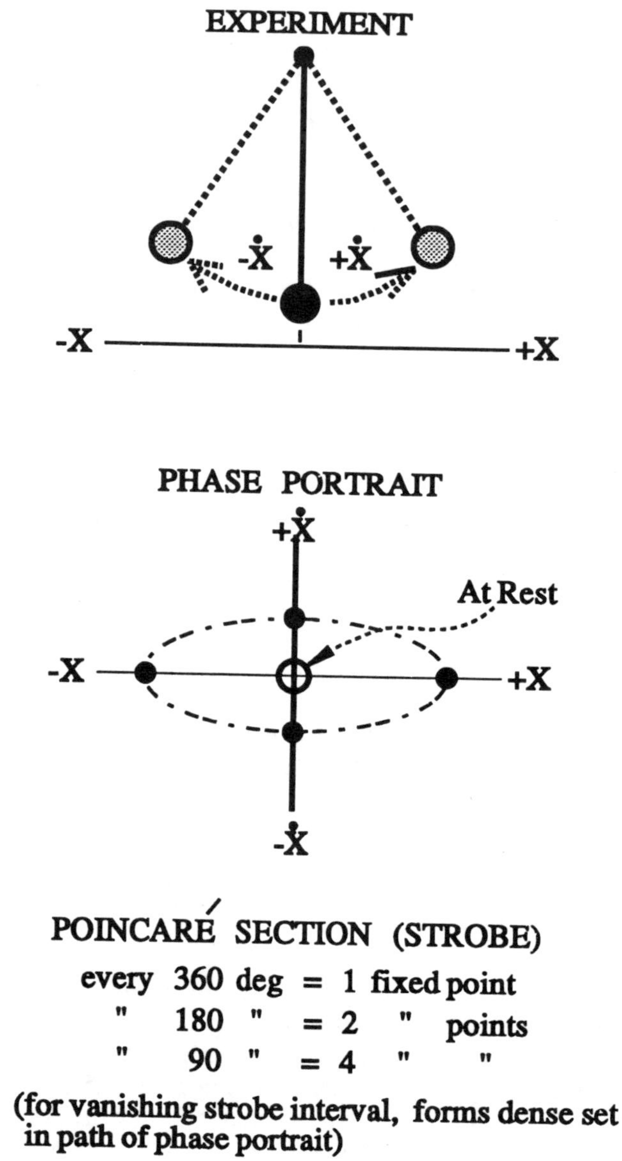

Figure 8.10: The simple pendulum and its phase portrait.

Trajectories of motion projected to the base trace out crudely elliptical paths that rotate around the vertical axis in complicated ways, as mentioned earlier (Miles, 1984, Fig. 8; Tritton, 1988, Fig. 24.4). Figure 8.12 shows caricatures resembling Poincaré traces of the X-Y motion on the base for a snapshot of a period-doubling attractor interval (left) and a chaotic attractor (right). Miles (1984) illustrates a wide variety of cases with examples of their power spectra, pointing out that some regimes correspond to strange attractors of the

8.19

SCHEMATIC NONLINEAR PENDULUM WITH DISSIPATION

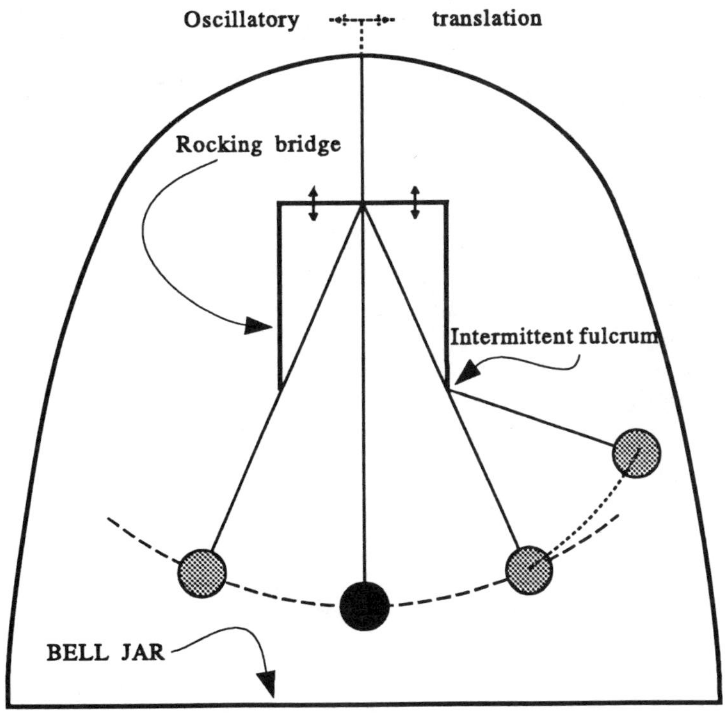

Figure 8.11: Hypothetical experiment in which a simple pendulum is subjected to nonlinear mechanical perturbations and thermal feedback.

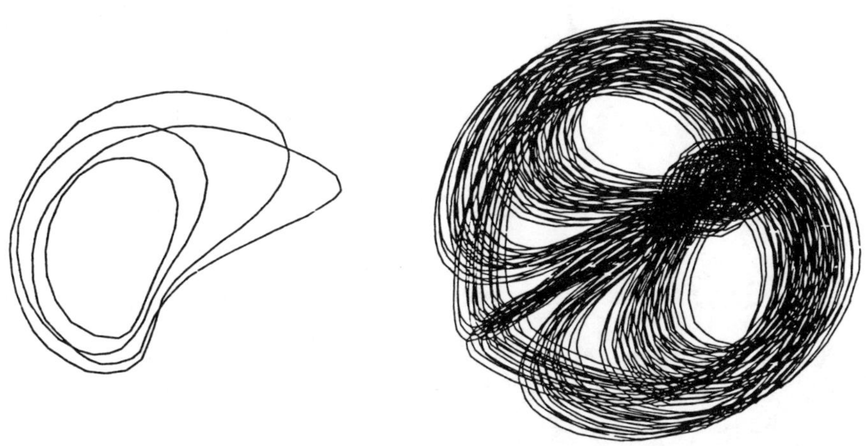

Figure 8.12: Schematic diagrams showing the general appearance of periodic (left) and chaotic (right) motions of the spherical pendulum, as expressed by Poincaré-like traces of the X-Y components of crudely elliptical paths on the base that rotate about the axis in complex ways (compare shapes of gyrations of the marker orbits in Fig. 8.5). The periodic orbits (in this case a portion of a period-doubling sequence) repeat the same trajectories, while the chaotic orbits diverge from previous cycles as they evolve, although tracing out the same general figure, the chaotic attractor. See Miles (1984) for details and documentation of periodic and chaotic regimes.

Lorenz type and Rössler type (cf., Figs. 8.24 and 8.29). Here is a case, then, where a simple mechanical device simulates chaotic attractors found in convecting fluids, with or without added effects of thermal feedback. However, frictional dissipation, as in the silicone experiments, may substitute for the externally imposed nonlinearities in a way that adds possibilities for self-organization of attractor regimes.

8.3.3 The cantilever beam.

Another system as simple as the pendulum is the elastic filament or beam caused to oscillate in a bistable manner between two fixed positions. The basic configuration is shown in Figure 8.13. Typical phase portraits and time series of bistable chaotic oscillations are shown in Figures 8.14 through 8.16. With weak forcing the beam can oscillate periodically much like the linear pendulum, and with strong asymmetric forcing it may lock on one or the other bistable positions, thereby illustrating a *fixed-point attractor* (note that a point attractor could be a static limit or a single point of a Poincaré map for a fixed oscillation frequency; see bottom of Fig. 8.10). For many other cases between these limits, however, complex patterns of bistable oscillation exist, such as the examples shown.

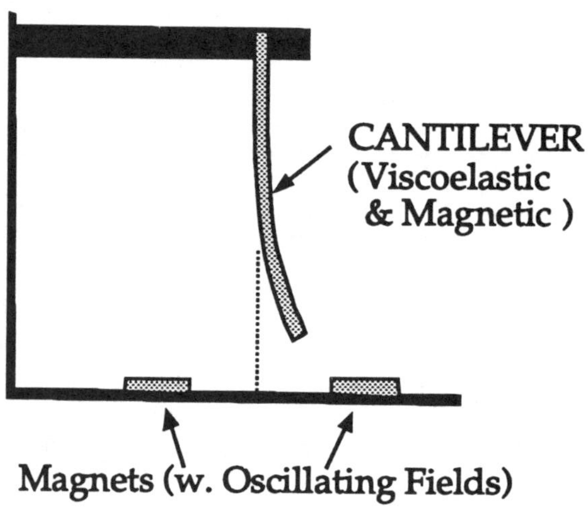

Figure 8.13: General configuration of vibrating cantilever beam and magnetoelastic ribbon.

Figure 8.14 is a typical phase portrait of a chaotic but relatively simple bistable oscillation. A Poincaré section of a similar case is shown in Figure 8.15, taken at some multiple of the forcing frequency. Note that, as shown in Figure 8.10, the Poincaré type of delay plot brings out that portion of a ciculation that is synchronous with the delay, as in viewing the motion with a stroboscope. For example, a stroboscopic meter focussed on the rim of a spinning circle with equally distributed fiducial marks (see Fig. 8.32) will 'see' only one mark if the flashes of the stroboscope are synchronized with an integer multiple of $2\pi f_0$ (where f_0 is the speed of rotation of the circle in cycles per sec, Hz) and will 'see' all marks if its frequency in Hz is synchronized with an integer multiple of the normalized spin frequency $(2\pi/p)f_0$, where p is the number of fiducial marks. However, note that this is not generally true if the marks are scribed by a nonlinear mechanism in which their positions are controlled by a mode-locked (rational) frequency, even though the unequal intervals are repeated exactly in every cycle. In that case, one has to either measure every position or measure the average frequency (see discussion of Fig. 8.32). The scatter of points in Figure 8.15 shows the chaotic deviations from a constant frequency and its harmonics. Figure 8.16 shows the time series of the bistability, which can be compared with that in Figure 8.2 suggesting that the two 'seismic' modes of the silicone experiment are analogous to a bistable flip from one pair of stress states (critical threshold and ground state) to another. These states are symbolized in Figure 8.2 by the dashed lines labeled T1, G1 and T2, G2.

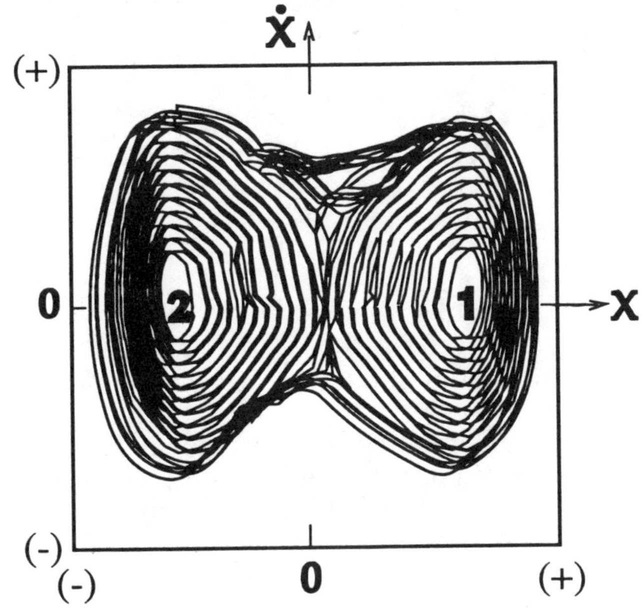

Figure 8.14: Schematic example of a chaotic attractor of a bistable vibrating cantilever beam. Many examples are illustrated and cited in Moon (1987).

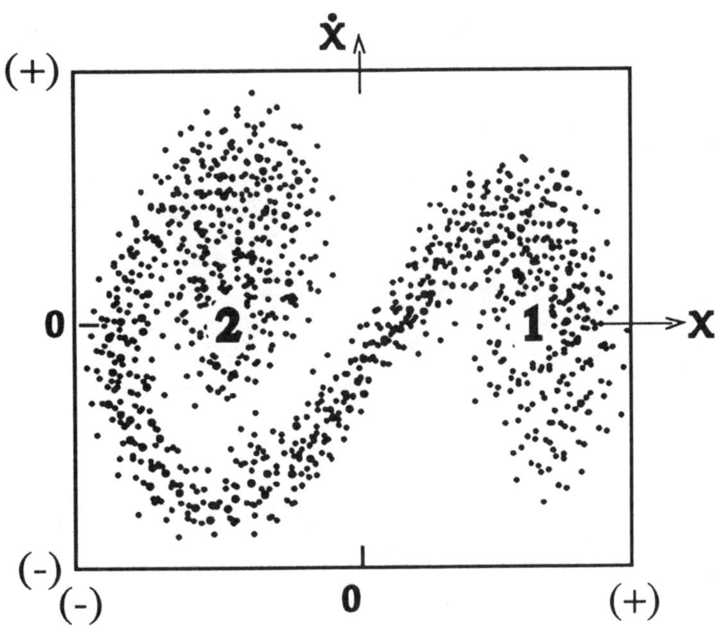

Figure 8.15: Schematic example of Poincaré section of chaotic bistable oscillator analogous to Fig. 8.14. See discussion in text.

8.22

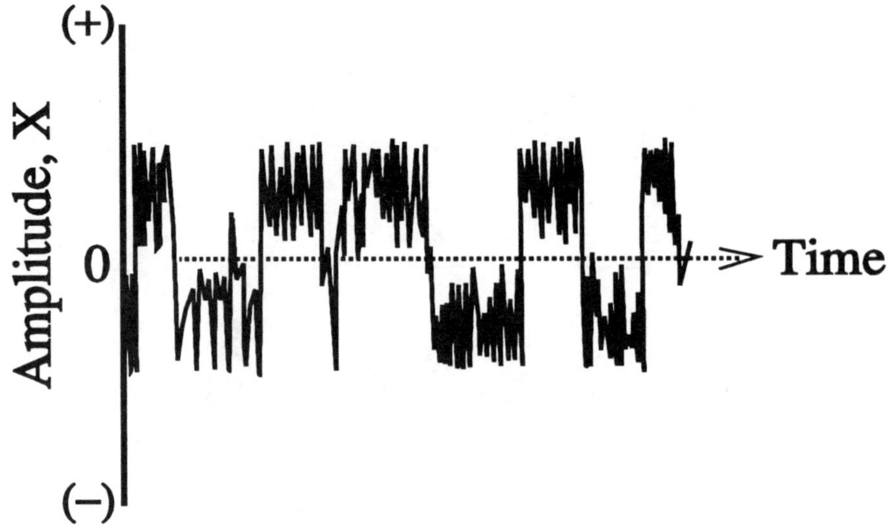

Figure 8.16: Schematic time series of bistable vibrating beam of Figs. 8.14 and 8.15.

A study of a vibrating magnetoelastic ribbon illustrated in Figures 8.17 and 8.18 suggests that in an asymmetrically stressed bistable attractor, when a shift occurs from the basin of the more stable attractor, there is a burst of chaotic oscillations of large amplitude. The critical stress state, or forcing frequency, where this occurs is termed an *interior crisis*. It is called 'interior' because it happens in the vicinity of an unstable periodic attractor embedded

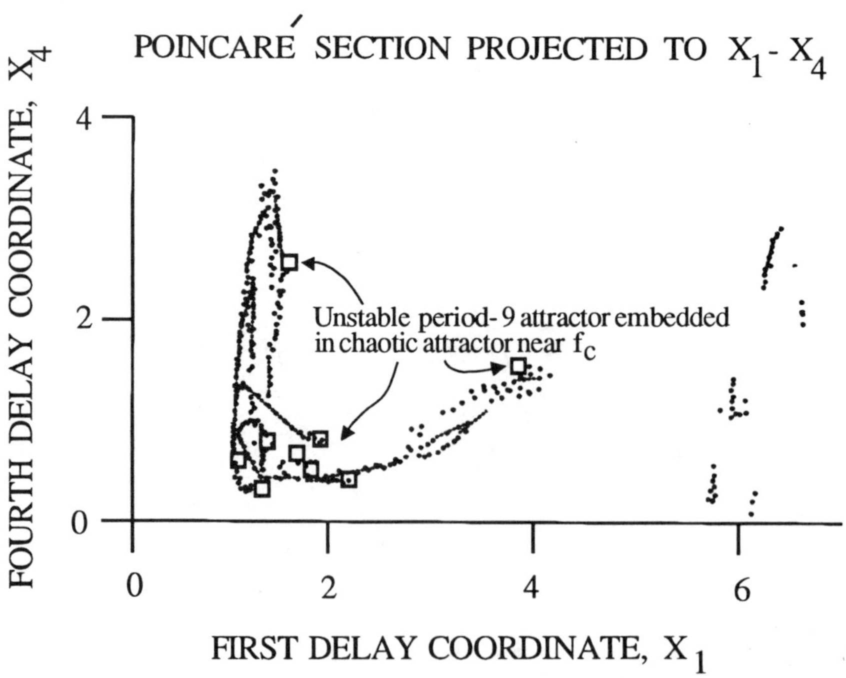

Figure 8.17: Poincaré section of a chaotic attractor of a magnetoelastic ribbon, schematically redrawn after Ditto et al. (1989). The squares mark the approximate positions of a 9-fold unstable periodic attractor embedded in the basin of attraction. Chaotic crises and bursting also exist in the chaotic region of circle maps and other coupled oscillators.

8.2 3

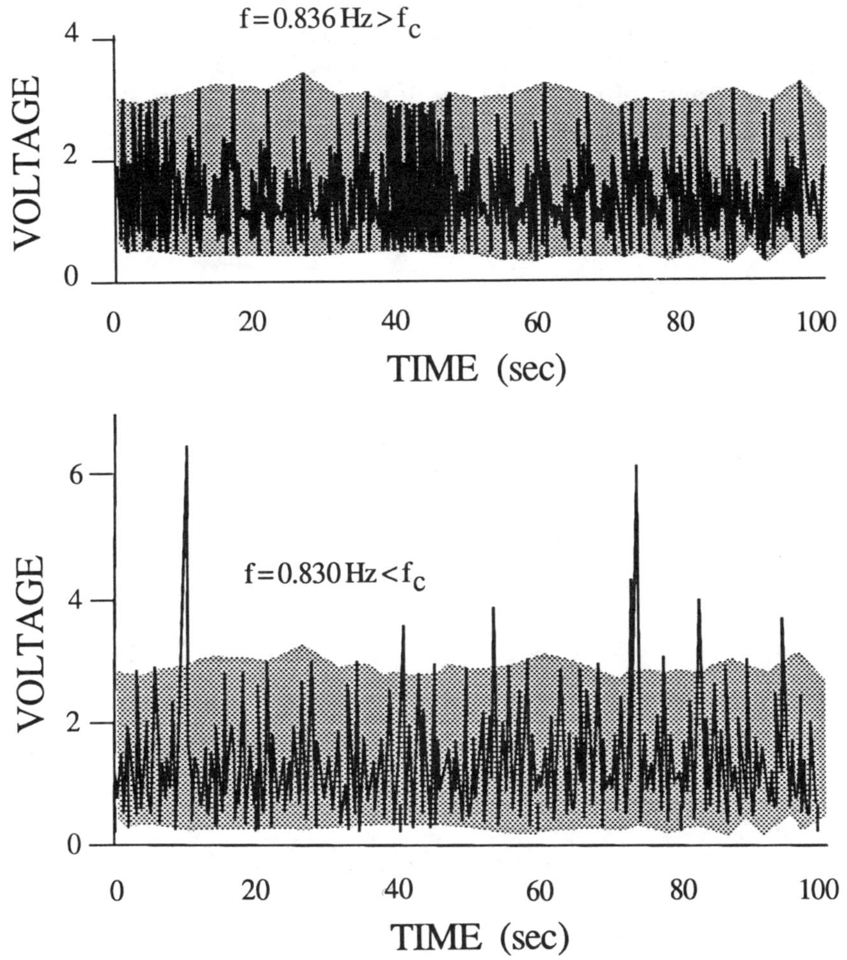

Figure 8.18: Time series of the chaotic attractor in Fig. 8.17, schematically modified and redrawn after Ditto et al. (1989). See discussion in text.

in the phase space of the chaotic attractor, as shown by the squares in Figure 8.17. Figure 8.18 shows the difference in time series at forcing frequencies a little above and below the critical frequency. The bistability is partly masked by the fact that the system was stressed asymmetrically, and the position readings are monotonic; the vertical is near X = 4, corresponding to the voltage output of the position monitor. Bursts to larger values of X are shown by the scattered points on the right in Figure 8.17 and by the higher voltages in Figure 8.18 (bottom). The shaded region is the chaotic attractor basin near the critical frequency. There is a remarkable resemblance to bursting phenomena in the silicone experiment (Fig. 8.2) and in the chaotic regime of circle maps. Note that this Poincaré map refers to a section in a higher-dimensional phase space.

It is evident that cantilever beams and ribbons could be insulated in such a way that flexural strain heating would be significant, so that thermal feedback would also influence the onset and characteristics of the chaotic modes, the onset of crises, and the time signatures of bistability. The magnetoelastic ribbon of Figure 8.17 might be especially sensitive to strain heating because of its large displacements, hence large work rates.

8.24

Attractor basins of bistable systems such as those discussed are shown in Figures 8.19 and 8.20. The former is of the simple type where there is a clear demarcation between trajectories that seek out one or other of the attracting vicinities (minima of the potential V in the top diagram of Fig. 8.19). In chaotic systems, especially where crises, bursting, and/or external noise may influence the motion, the trajectories of attraction are not as simply differentiated. In such cases the structure of the attractor basin is itself fractal, and

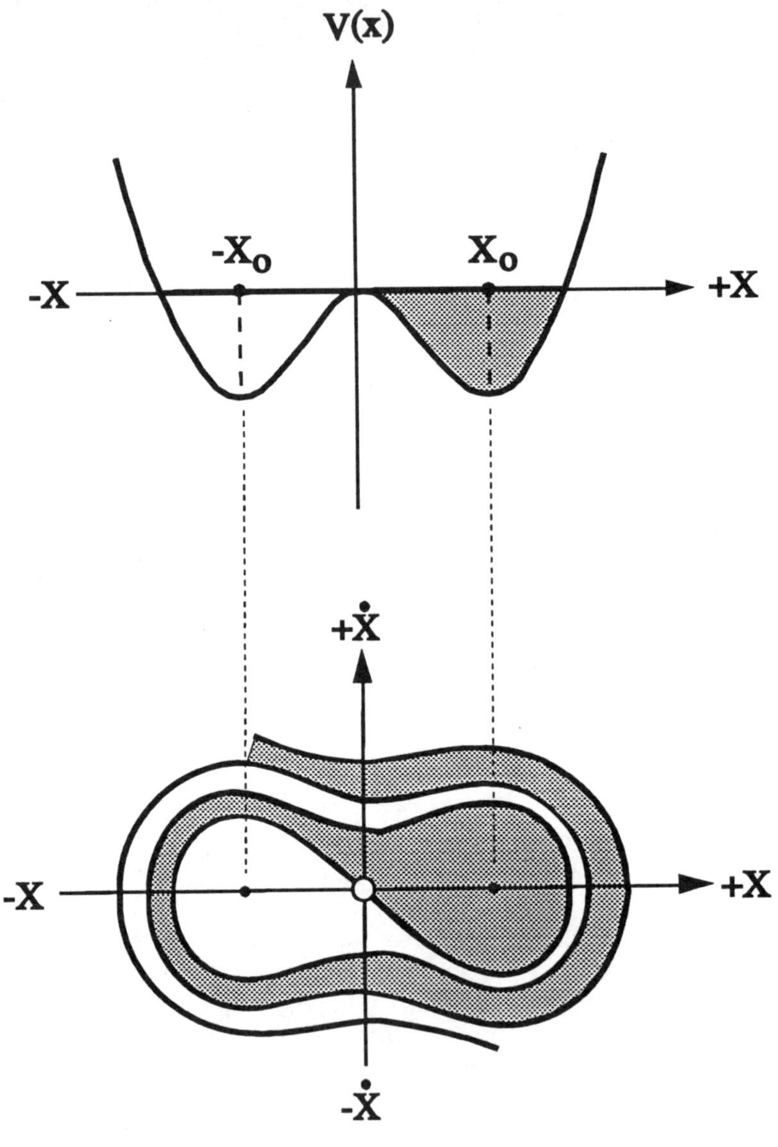

Figure 8.19: Section through a two-well potential surface (top), where V(x) is the potential, and the white and grey regions delineate the wells of two fixed points of a particle that comes to rest at one or the other minimum. Like a two-slot roulette wheel, the possible paths of a ball will go to one or the other minimum uniquely and smoothly as a function of initial velocity, momentum, etc., as shown (bottom). With nonlinear interactions, however, the paths are nonunique, and the attractor basins (white or grey) are no longer smooth, nor necessarily even continuous, as seen in Fig. 8.20.

it looks somewhat like those of the phase portraits of strange atttractors. Figure 8.20 is a caricature of a demonstration by Moon and Li (1985) of an incredibly rich fractal structure found to exist between the periodic and chaotic regimes of a bistable oscillator ('two-well potential' attractor). In this case the fractal structure represents the probability of whether a given initial state will be drawn to the vicinity of one or the other bistable loci. Although the stylized sketch of the fractal structure has turned it into a *connected fractal*, there are no unique boundaries between the light and dark areas. This has powerful implications for questions of the predictability of even relatively simple bistable systems and forms a connection with problems of fractal intermittency in chaotic systems, such as that of Figure 8.2, volcanic tremor (Shaw and Chouet, 1989), and many regimes of the circle map (e.g., Jensen et al., 1985; Halsey et al., 1986). The fractal dimension of the basin 'boundariy', as portrayed by Moon and Li (1985) depends on the uncertainty of initial conditions, and was not given explicitly. However, viewed as a 'coastline' problem in Figure 8.20, the *capacity fractal dimension,* for *topologic dimension* 2 in phase space, is roughly 1.2 to 1.3 (cf., Moon, 1987, p. 250 ff.).

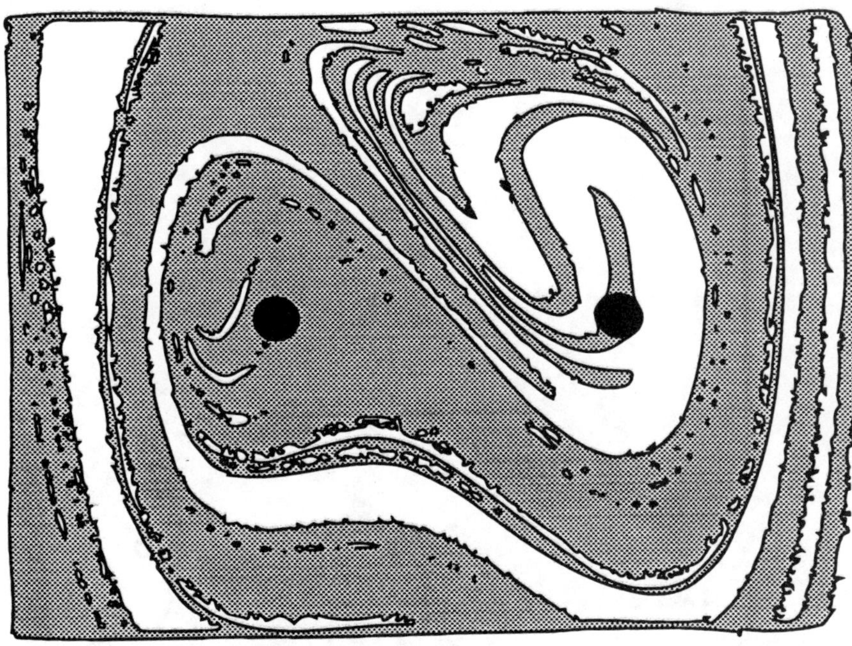

Figure 8.20: Caricature of the fractal basin boundaries of a nonlinear two-well potential (bistable oscillator) that is neither simply periodic nor strictly in the chaotic domain (e.g., this could happen near the divide of Fig. 8.19 where a small perturbation can change the outcome even if the resulting trajectory might subsequently follow a periodic orbit in the vicinity of one or other of the potential wells); the drawing is grossly stylized compared to the delicate fractal tracery illustrated by Moon and Li (1985).

8.4 Convection Loops

In this section I broadly include the Lorenz system and double-diffusive (thermosolutal) convection because they have a number of regimes in common. The simple convection loop also resembles the nonlinear pendulum and the bistable oscillator in its behavior.

8.4.1 The simple loop.

The motion of a fluid around a circular or oblong closed tube heated at the bottom and cooled at the top might be considered to be the prototypical magma chamber. Furthermore, the flow has a number of properties that have already been seen in previous illustrations. Figure 8.21 is the simplified model, modified from Welander (1967), where the graph at the bottom shows the initial stages of the cycle.

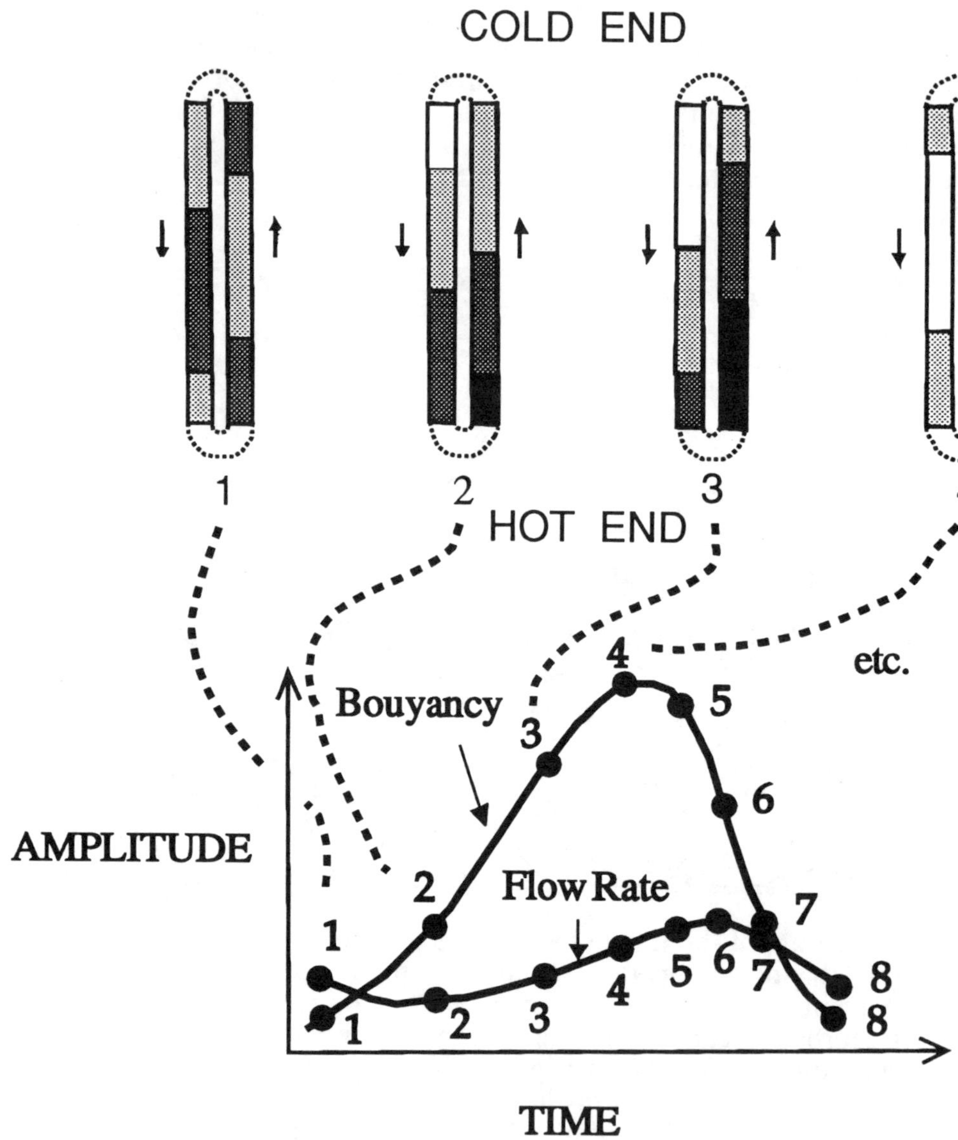

Figure 8.21: Schematic of the simple convection loop, modified and redrawn after Welander (1967, Figs. 10 and 11). The graph shows the growing bouyancy of the heated (black) portion of the fluid in a partial cycle.

It is obviously impossible in Figure 8.21 to tell which way the flow will cycle, and in this regard the problem resembles the two-well potential with fractal basins of attraction (cf., caption of Fig. 8.20). In Figure 8.22, schematic time series are shown from an analogous study by Creveling et al. (1975) for steady, oscillatory steady, and oscillatory reversing flow regimes. Figure 8.23 shows a case of chaotically reversing flow in which feedback control was used to mediate the chaos, negate reversals, and control the amplitude of oscillations. In that study, by Singer et al. (1991), the temperature signals at the 3 o'clock and 9 o'clock positions activated a servo mechanism that opposed the growth of the instabilities of either sign. The result resembles the suppressed oscillations in the lower graph of Figure 8.2. In that case there was a natural servo mechanism by which the tendency for vapor bubbles to grow in the sheared and thermally perturbed field were countered by the rate at which they decayed and were resorbed into the circulating flow.

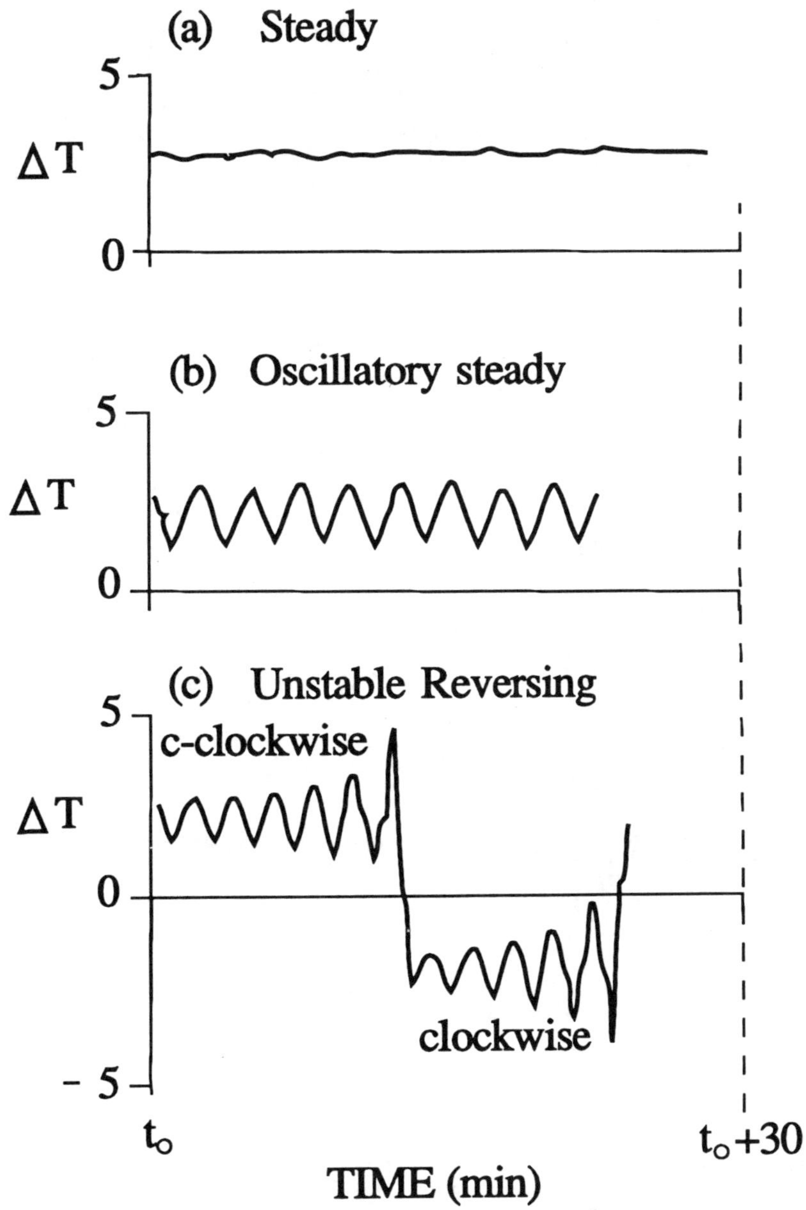

Figure 8.22: Time history of different regimes of the simple convection loop, schematically modified after Creveling et al. (1975).

8.28

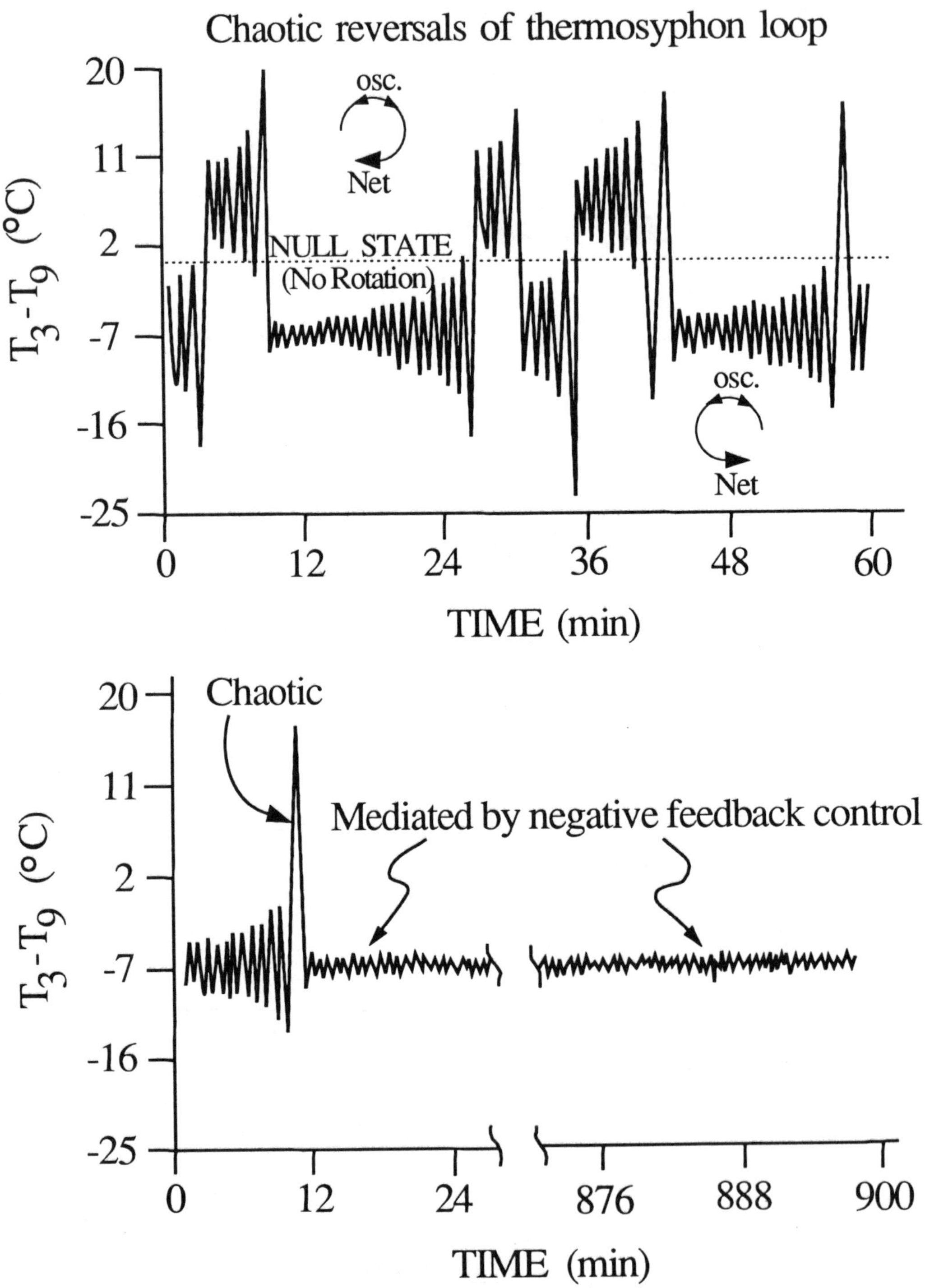

Figure 8.23: Chaotic bistable states of the simple convection loop and 'controlled' chaos, schematically modified after Singer et al. (1991). See discussion in text, and compare with Fig. 8.2 (bottom).

8.4.2 <u>Convection loops of the Lorenz and double-diffusive types.</u>

Reference is often made in the literature to the "Lorenz Attractor", as though it is only one unique object. The system studied by Lorenz (1963) has many regimes, including periodic cycles that are virtually identical with the convection loop (e.g., Tritton, 1988, p. 394 ff.). In a sense, *tuning* is the central issue in nonlinear dynamics. Attractor states are tuned, sometimes robustly and sometimes with excruciating sensitivity, by the relationships between forcing and the strength of coupling among parameters of the system, such as its natural frequency modes and/or those which may be imposed on or induced in the system by external (e.g., heating) or internal (e.g., thermal feedback) mechanisms. It is easy to fall into the habit of speaking of this or that 'attractor' as though it were characteristic of a particular system. As this discussion indicates, even the simplest systems potentially have an infinite number of attractor states, a fact that study of the circle map, an algorithm in one dimension and with only two adjustable parameters, makes abundantly clear if it was not already obvious from studies of quadratic maps (e.g., May, 1976; Feigenbaum, 1980). Two examples of Lorenz attractors are shown in Figure 8.24. On the left is a snapshot of one attractor state among a sequence of period-doubling states (analogous to that of the nonlinear pendulum in Fig. 8.12), while on the right is a portion of the famous 'butterfly' strange attractor (see Sparrow, 1982; Tritton, 1988).

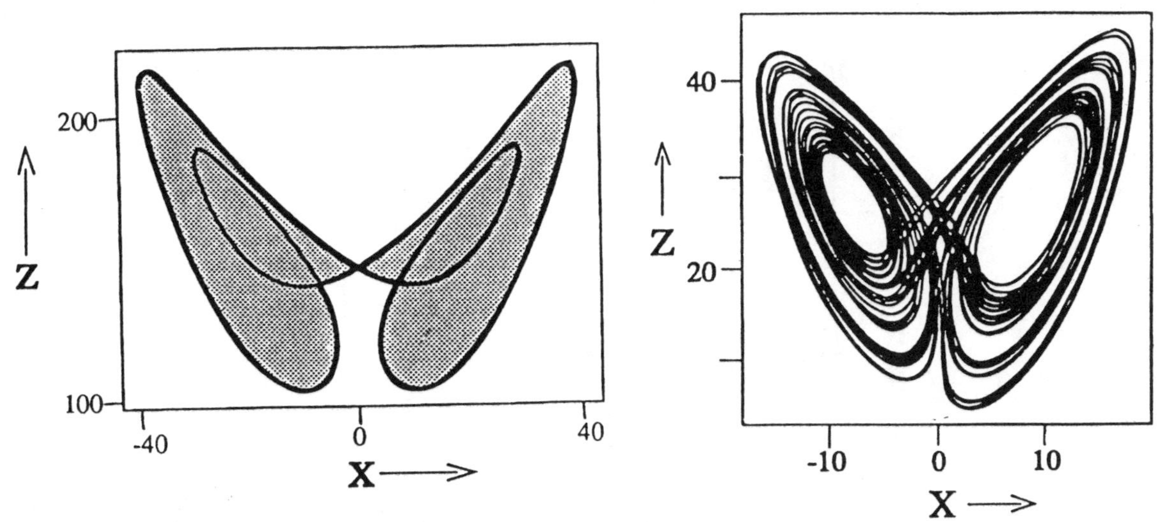

Figure 8.24: A portion of a period-doubling sequence (left) and a strange attractor (right) of the Lorenz equations representing a simplified mathematical model of cyclical convection, schematically redrawn from examples in Sparrow (1982). Equations and representation as a waterwheel also are described in Sparrow (1982, Appendix B), and the relationship to the simple convection loop is demonstrated in Tritton (1988, p. 246 ff.).

Although the cases differ topologically, the circulation between the two different foci of the Lorenz phase portraits resembles the bimodal patterns of bistable oscillators (e.g., Fig. 8.14). In the Lorenz strange attractor, however, the trajectories cross and mix the flow in complex ways due to sensitive divergence (e.g., resembling the case of Fig. 8.4, where in three dimensions the marker also circulates in the mirror image of Fig. 8.3). An interesting demonstration is given by Crutchfield et al.(1986) where a 'digital dye patch' is allowed to mix in the phase space of the Lorenz strange attractor, much as if I had inserted a cluster of many particles in the silicone experiment instead of using only two particles to test the divergence because I didn't want to lose track of the specific markers (see discussion of Fig. 8.4). Their result vividly demonstrates sensitive dependence on initial conditions as well as phenomena of folding and kneading (somewhat like stages in the process of making Samurai swords or rolled biscuit dough).

The case of double-diffusive convection is included because it makes connections with all of the experiments discussed and with interesting doubly-terminating bifurcation sets (forward-reverse bifurcation "bubbles" in the terminology of Moore et al., 1983) and other phenomena seen in the chaotic regimes of circle maps. Moore et al. (1983, Fig. 1) and Knobloch et al. (1986, Figs. 5 and 8) show resemblances of their mathematical solutions to regimes of the Lorenz equations. They also describe histories of growth of attractors, showing time series of coupled mass and heat transport as well as phase portraits and power spectra of the frequency structures. Huppert and Moore (1976) illustrate the growth of oscillatory instabilities analogous to those shown earlier in the silicone experiment, except that in their case the oscillations involve both heat and solute concentration. Huppert and Turner (1981) describe the general phenomenology of double-diffusive convection with applications (an obvious candidate being the transport regimes of magma chambers; cf., Shaw, 1974). Some of these effects are recapitulated in Figures 8.25 through 8.27.

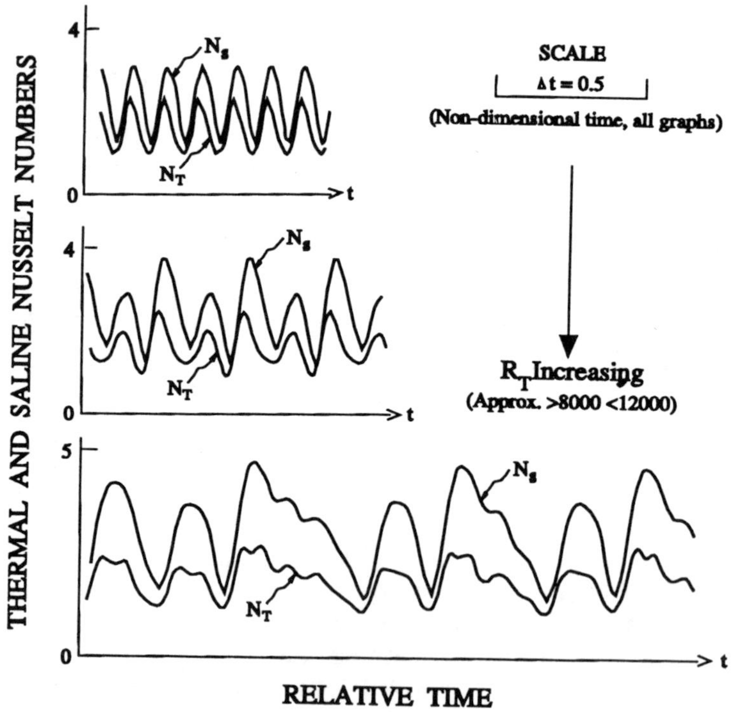

Figure 8.25: The growth of oscillatory instabilities of double-diffusive convection, schematically redrawn from Huppert and Moore (1976, Fig. 11). N_T is the thermal Nusselt number and N_S that of salt, both defined in an analogous manner to that of Figure 8.9 and over a similar range of thermal Rayleigh numbers, R_T. The increasing complexity downward is evident (cf., chaotic time series of the silicone experiment in Fig. 8.7).

8.31

In Figure 8.26 the time series of a stronger oscillation hints at how chaotic mixing comes about. Note that the curves of heat and mass transfer that were roughly synchronous in Figure 8.25 are becoming increasingly out-of-phase, meaning that there is a competition between the thermal and saline modes of transfer. This effect resembles that of coupled oscillators in which the resulting frequency regimes depend on the nature of resonances (periodic or chaotic) between the modes of oscillation. Thus, we would expect to see some of the same bifurcation phenomena, eventually including interior crises and chaotic bursting, evident in regimes of the circle map (e.g., like those of the magnetoelastic ribbon in Figs. 8.17 and 8.18).

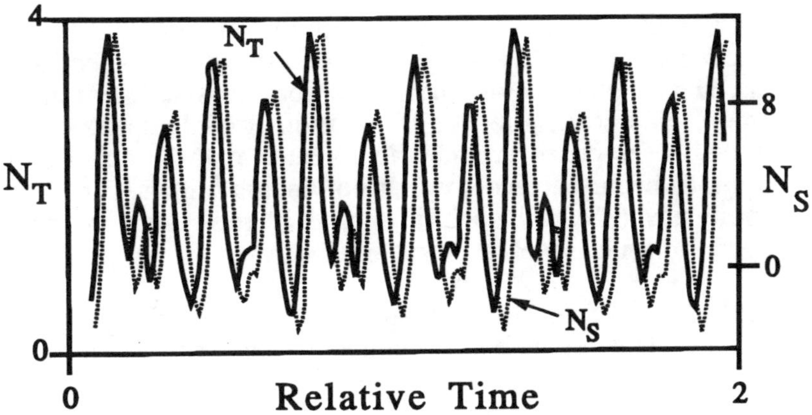

Figure 8.26: Time series of a period-doubling regime that may eventually lead to chaotic oscillations in double-diffusive convection, schematically modified and redrawn after Knobloch et al. (1986, Fig. 8). Compare with discussion of Fig. 8.27.

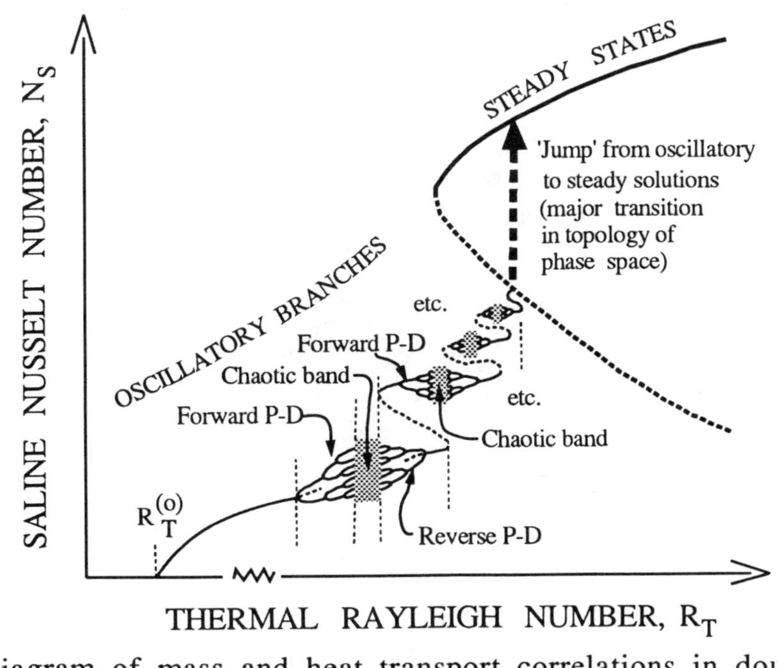

Figure 8.27: Diagram of mass and heat transport correlations in double-diffusive convection analogous to Fig. 8.9 for the silicone experiment, schematically redrawn and modified from Knobloch et al. (1986, Fig. 6); symbol P-D refers to period-doubling regimes.

8.3 2

These effects are summarized in Figure 8.27 where the saline Nusselt number, N_S, is plotted against the thermal Rayleigh number, R_T, in the same manner as N_T versus R_T in the silicone experiment of Figure 8.9, coincidently beginning at about the same value of R_T. The nature of the period-doubling and chaotic intervals (the "bifurcation bubbles" of Moore et al., 1983) will be seen in more detail in diagrams of the dripping faucet and of circle maps.

8.5 Dripping Faucets and Circle Maps

These two types of nonlinear systems have so much in common, that I lump them together in this section. However, it should also be evident, almost by inspection now, that all of the systems discussed have many common aspects from the standpoint of potential periodic and chaotic regimes. Therefore, I let the illustrations mainly speak for themselves, and conclude the review in the next section with selected natural systems that in my view illustrate the ubiquitous occurrences of similar nonlinear regimes. The latter examples are drawn from geophysics (broadly speaking) and astrophysics, representing extreme differences in space and time as well as in material states.

To set the stage, Figure 8.28 illustrates an analog model of a dripping faucet represented as a coupled oscillator. This brings out its connection with circle maps. One oscillation is

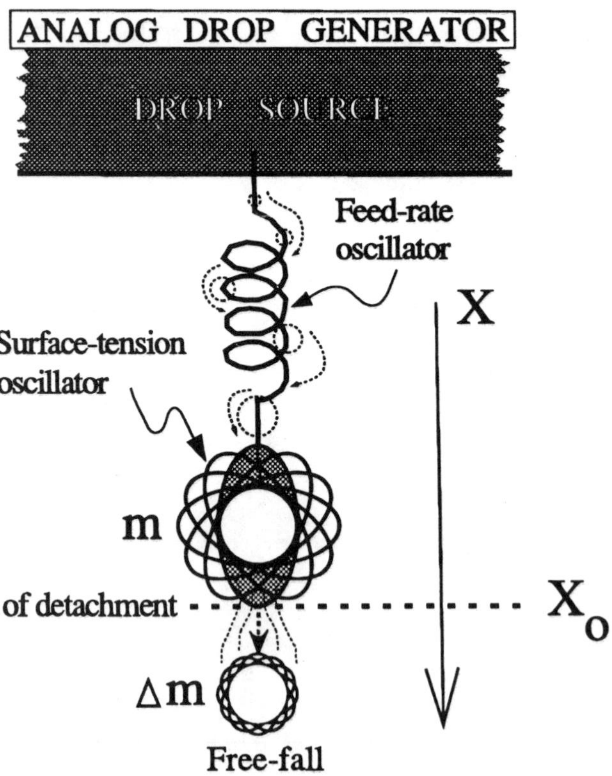

Figure 8.28: Hypothetical example of a dripping faucet expressed as a coupled oscillator, in part suggested by the analog model of Shaw (1984, Fig. 9) and the experiments of Wu and Schelly (1989).

that of the feed rate as a new drop begins to form, while the other is the changing oscillatory modes of the drop as it approaches the state of detachment to form a freely falling drop. Shaw (1984) analyzed such a model in the form of a spring and a mass, part of which detached at a particular position in the motion of the spring, resulting in the chaotic attractor shown in Figure 8.29 and the time series of Figure 8.30. Resemblances to circulations in the phase plots of the silicone experiment may be noted.

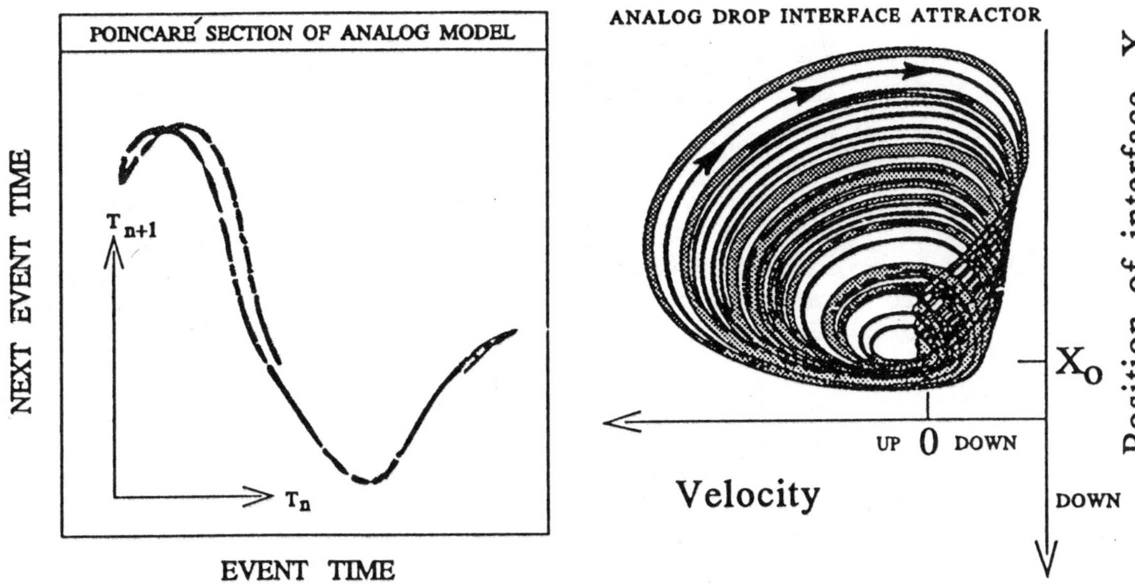

Figure 8.29: A strange attractor (right), schematically redrawn from the analog model of Shaw (1984, Fig. 11), that resembles the Rössler-type of strange attractor (e.g., Holden and Muhamad, 1986, p. 23 ff.) also demonstrated by chemical reaction oscillators. A Poincaré section near X_o is shown on the left (e.g., Shaw, 1984, Fig.10). Compare with circulations in plots of the silicone experiment (Figs. 8.3 to 8.6).

The richness of the dripping faucet model as a source of periodic and chaotic oscillations is summarized in Figure 8.31, which is a caricature from the experiments by Wu and Schelly (1989). The figure above the drawing is an arbitrary interval of the generally chaotic regime of the circle map in which the alternating period-doubling and chaotic intervals, analogous to the "bifurcation bubbles" of Moore et al. (1983), are abundantly displayed. Note the simplified but strong resemblance to the dripping faucet regime in the left half of the diagram below it. Such diagrams are called *bifurcation maps* or *tuning maps*, because they illustrate in a concise way the diversity of possible attractor states as a function of one or more driving parameters, in this case the supply rate of drop-forming solution (analogous tuning of bifurcation states of volcanic cycles, expressed in terms of the supply rate of magma, are illustrated and discussed in Shaw, 1987a , 1988, although similarity to the dripping faucet was not explicitly invoked there).

8.34

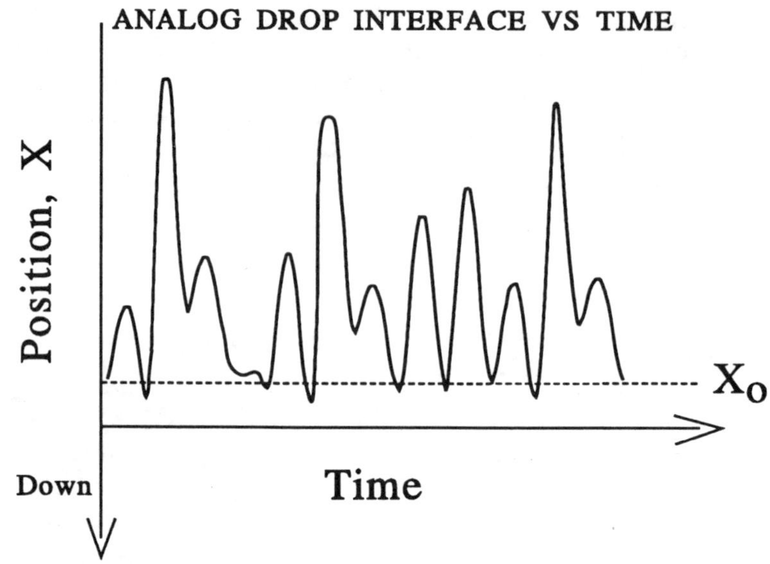

Figure 8.30: Example of time series of the analog model of the dripping faucet in the strange attractor regime of Fig. 8.29 (schematically redrawn from Shaw, 1984, Fig. 12). Compare with previous time series, especially Fig. 8.7.

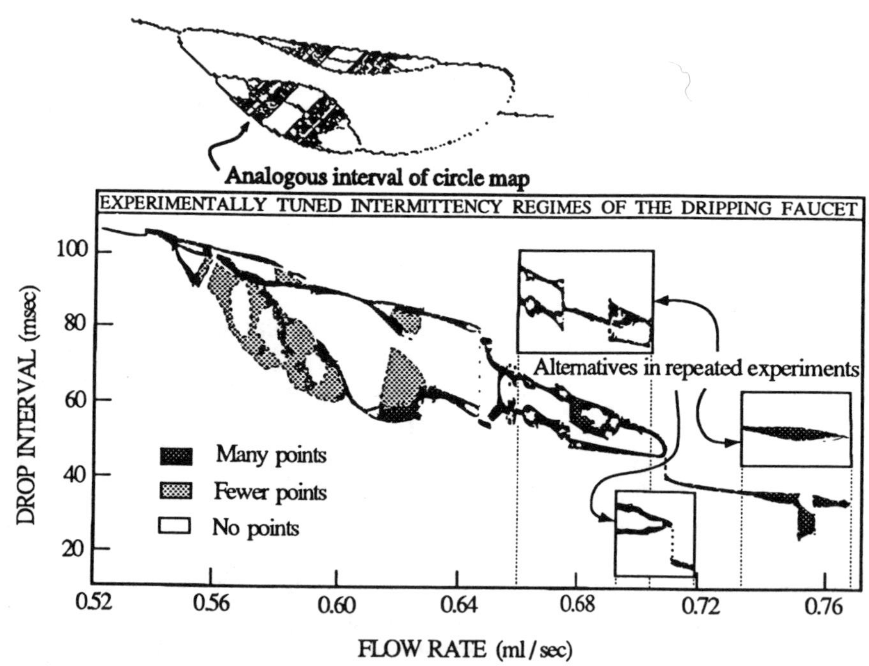

Figure 8.31: Caricature of the dripping faucet regimes, with surfactant, as studied experimentally by Wu and Schelly (1989, Fig.4), which they referred to as a "dripping spectrum" and is called a *bifurcation map* or *tuning map* here. The surfactant reduces surface tension, thereby sharpening the tuning of the drop intervals. This effect is analogous to changing viscoelastic compliances in the cantilever beam or magnetoelastic ribbon (Figs., 8.13 to 8.18). The bifurcation pattern shown above the diagram is a portion of a circle map regime computed by the author (cf., Fig. 8.35).

8.35

The illustrated regimes of the dripping faucet bring richer meaning to the idea of 'bistability'. Note that there are many values of flow rate for which two or more values of drop interval coexist. These clearly represent conditions under which there may be bistable or multistable states, choices among which may be uncertain in the sense of the fractal basins of attraction already discussed. This situation is underscored by the insets within the drawing, which are alternative attractor states found in repeated experiments.

The next sequence of illustrations portrays some regimes I have computed using an algorithm for the sine circle equation, as explained in Shaw (1987b):

$$X_{n+1} = X_n + \omega - (K/2\pi)\cdot\sin(2\pi X_n), \tag{8.1}$$

where the strength of coupling is represented by K, and the phase difference or *bias* of the coupling frequency is given by the fractional angle ω with values between 0 and 1.

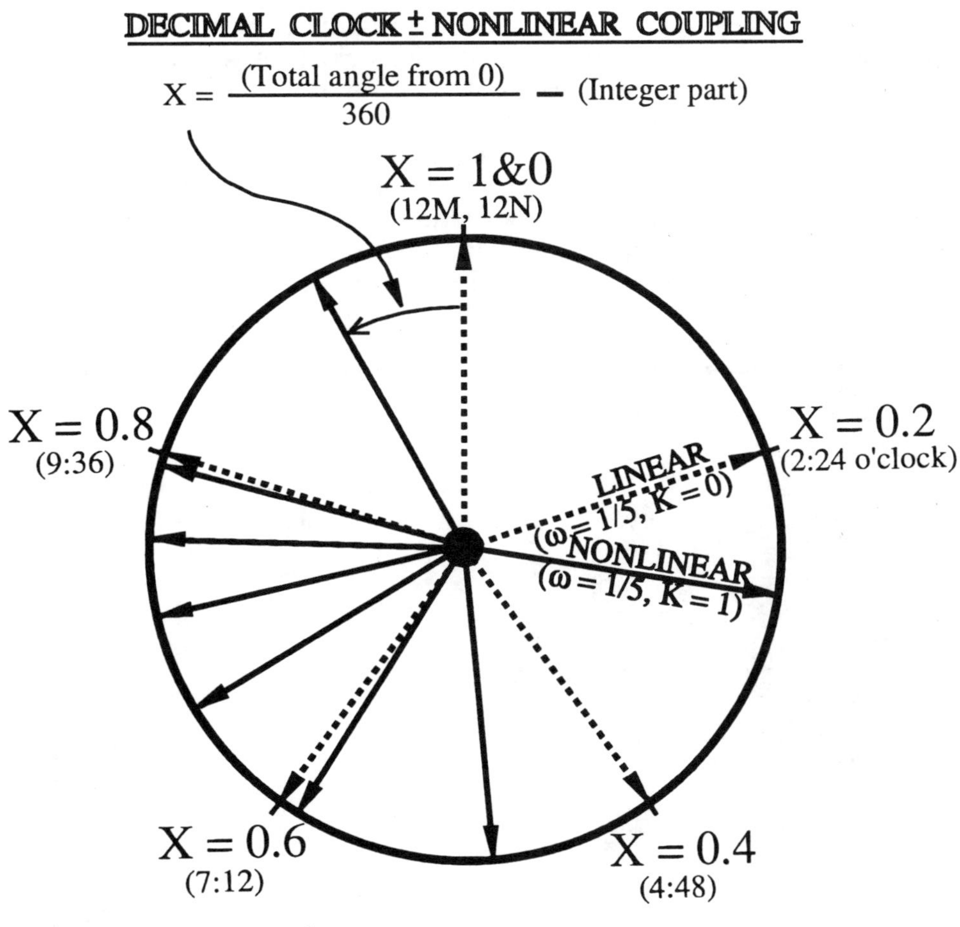

8 nonlinear time positions per cycle = 5 linear ones in this example.
The average nonlinear time increment is 1.5 hrs (12-hr clock), but are unequal;
Actual increments at this nonlinear mode-locked setting range from 0.5 to 4.5 hrs.

Figure 8.32: The circle map as 'nonlinear clock' and paradigm for coupled nonlinear oscillators, from studies by the author (cf., Shaw, 1987b, Fig. 1).

8.36

The angular position variable X (where the subscripts n and n+1 represent its present and subsequent value, respectively) is expressed as a decimal fraction of a full cycle (i.e., continued rotation is accounted for by subtracting the integer part of the cumulative angular rotation). The presence or absence of linear or nonlinear resonance conditions are easily observed as periodic and/or aperiodic distributions of the X-values. The 'natural' frequency is just the rotation of a point around the circle with zero coupling and bias. The average number of iterations about the circular cycle is expressed by the so-called winding number, W, given by:

$$W = \text{Limit (large N) of (Total Angular Rotation from Initial Angle, } X_o)/ N \qquad (8.2)$$

where N is the total number of iterations of X for a given setting of K and ω. Theoretically, N must be infinite to give an accurate value of W, but in practice it is about a thousand. These relationships are shown in Figure 8.32 for a simple case of mode-locked resonance.

Space does not permit discussion of all possible regimes, which are outlined in Shaw (1987b) and references cited there. Periodic solutions exist for $K < 1$, while both chaotic and complex period-doubling windows exist for $K > 1$. At $K = 1$ there is a fractal distribution of all possible mode-locked (rational) frequencies as well as quasiperiodic (irrational) frequencies one of which, for W equal to the inverse of the Golden Mean, is now famous as the paradigm for quasiperiodic routes to chaos in coupled oscillations. This is the case referred to earlier in studies of convection (Jensen et al, 1985; Fein et al., 1985), electronic oscillators (Gwinn and Westervelt, 1987), and seismic tremor (Shaw and Chouet, 1989). Examples of circle maps are given in Figures 8.33 through 8.35.

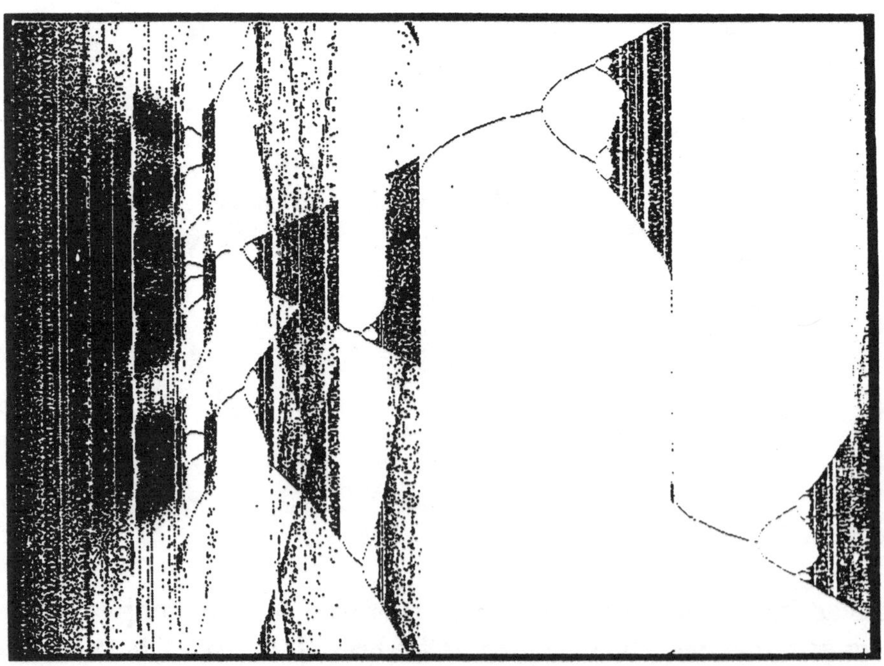

Figure 8.33: Bifurcation map of sine circle function for values of K (horizontal) between 0 (left) and 5 (right) for ω ~ 0.61, giving W ~ 0.6180339.., *inverse of Golden Mean,* at K = 1 (cf., Fig. 8.36). Vertical axis is X (0 at bottom, 1 at top).

8.37

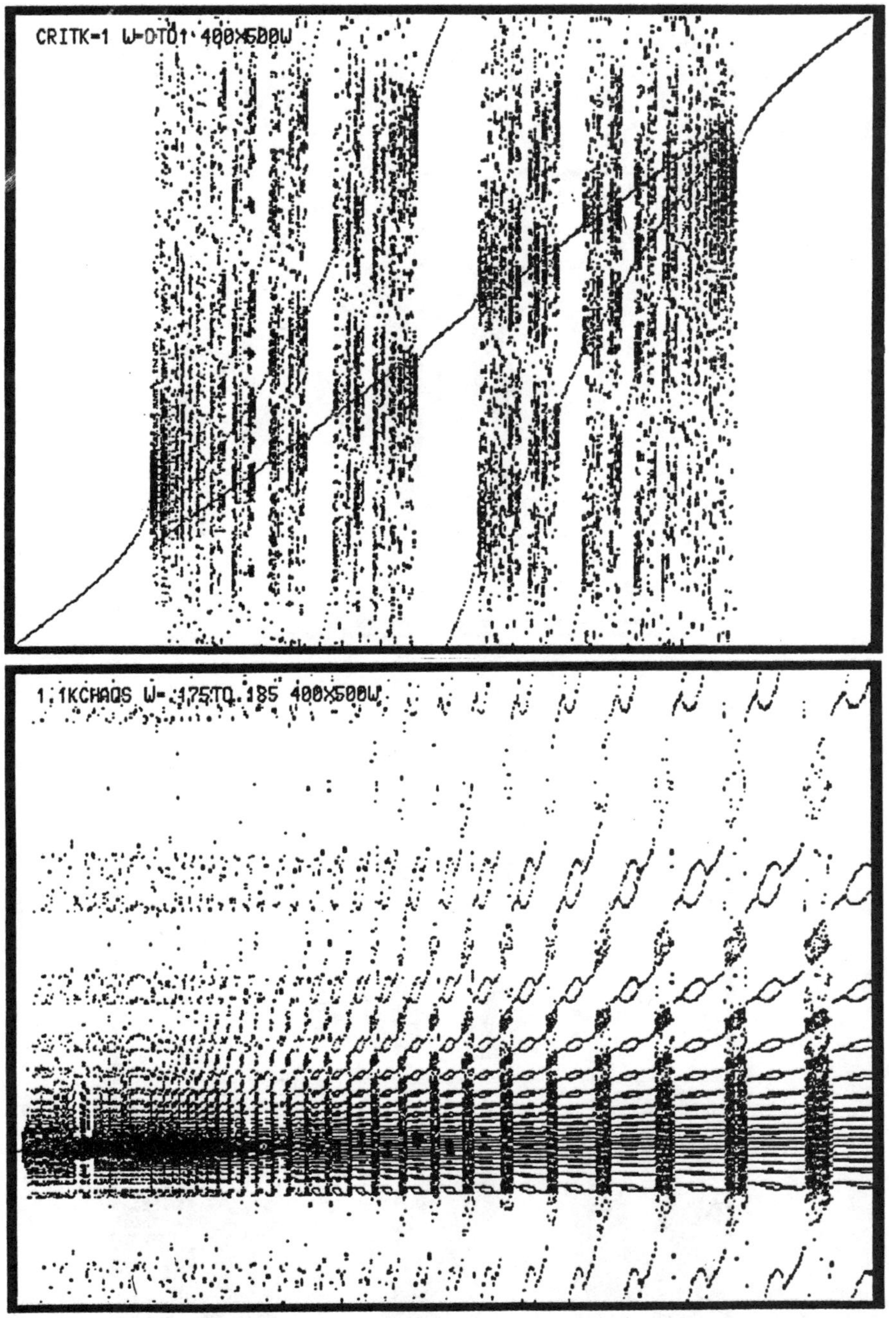

Figure 8.34: Bifurcation map of sine circle function for K = 1 and with both X (vertical) and ω (horizontal) between 0 and 1 (top) representing a complete cycle of mode-locked and quasiperiodic states; and for K = 1.1 (bottom) portraying a smaller region of ω near the initial appearance of multiple periods at left of upper diagram.

Figure 8.33 represents a bifurcation map as a function of increasing K from 0 to 5, representing high degrees of nonlinear coupling above K = 1 (note extreme jumps in period-doubling and chaotic streaks), while Figure 8.34 gives plots of X versus ω for the values K = 1 and K = 1.1, respectively, the critical value and a value just into the chaotic regime where cascades of period-doubling "bubbles" are seen to be forming (bottom diagram). Figure 8.35 shows small regions in the chaotic regime where "bifurcation bubbles" (or *bifurcation synapses*, as I prefer to call them) divide and/or merge in complex ways reminiscent of neural phenomena, dripping faucet regimes (Fig. 8.31), and double-diffusive oscillatory branches (Fig. 8.27).

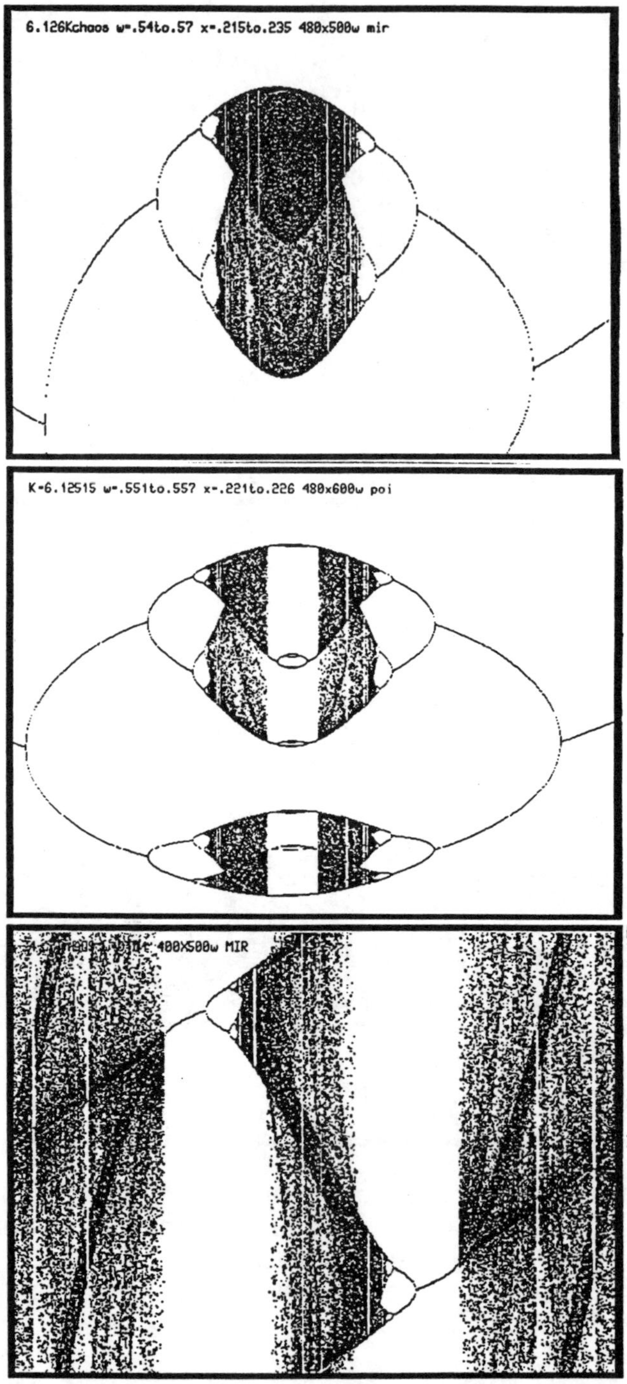

Figure 8.35: Bifurcation regions of the circle map at supercritical values of K >> 1 where 'synapses' (or "bubbles") are richly displayed.

8.39

The behavior of the winding number, W, in the supercritical region elucidates the multistability issue already identified in earlier systems and the dripping faucet map of Figure 8.31. Figure 8.36 shows regions of two or more coexisting values that may jump about abruptly at high values of K. Remember that W represents an 'average period' at a given setting of ω and K, while the drop interval of Figure 8.31 is like reading W directly on a gauge (see discussion of experimental method in Wu and Schelly, 1989). Both, however, depend on how many iterations, or drops, are averaged. This aspect of the supercritical sine circle map also has been studied by Ding and Hemmer (1988).

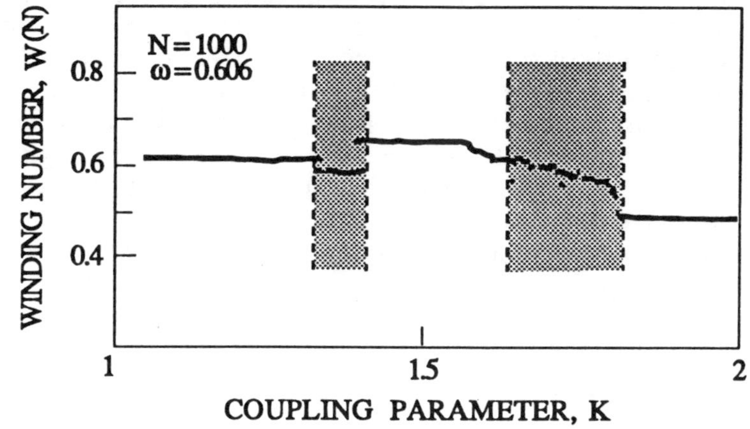

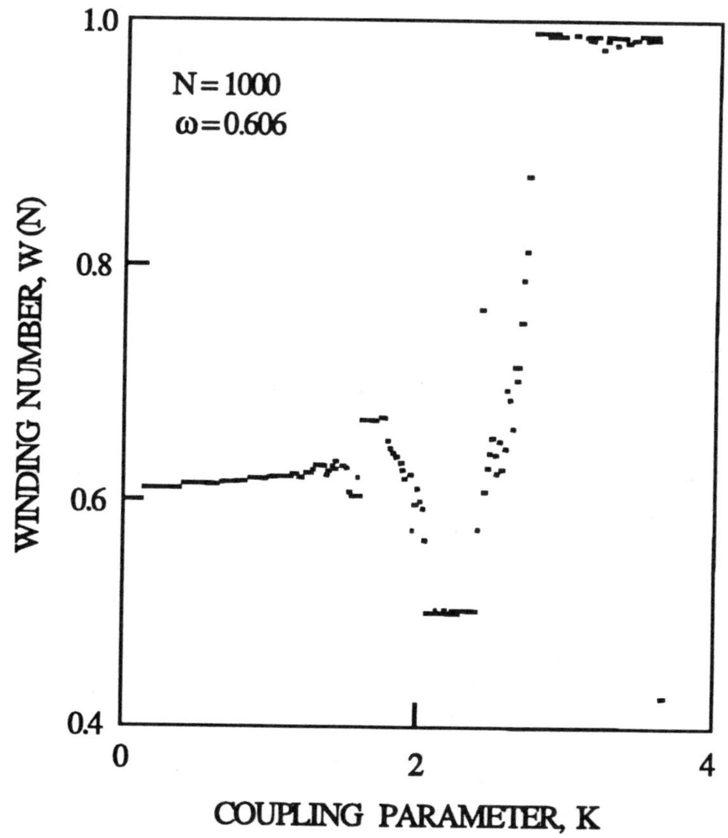

Figure 8.36: Winding numbers as functions of supercritical K illustrating fractal multistability regions in the circle map. See Eqs. (1) and (2).

8.6 Conclusion: Selected and Conjectural Geophysical Applications

In concluding this review of nonlinear case histories, I illustrate three additional cases of widely different time and spatial scales which resemble aspects of the examples discussed. One concerns terrestrial volcanic phenomena, another chaotic asteroid and planetary orbits, and the third the galactic setting of the Solar System. Another case study, mentioned in the introduction, applied the dripping faucet model to the formation of plasmoids in the Earth's magnetotail (Hones, 1979; Baker et al., 1990). The diagrams, when viewed in terms of the array of nonlinear time series and their attractors already displayed, pretty much explain themselves, or at least raise interesting questions concerning the prevalence of chaotic dynamics even with regard to effects usually considered to be precisely ordered (like planetary orbits) and/or exogenously remote (like the dynamics of the Milky Way Galaxy in which all the other attractors are embedded). In fact, I conclude the discussion here and leave it to the previous figures, to notes in the captions, and to the imagination as to the implications of the following figures, recalling that thermomechanical coupling, especially in the contexts illustrated at the beginning of this tour, is not just the whim of a few experimentalists like Rumford and Joule but is at the heart of all *Motion*.

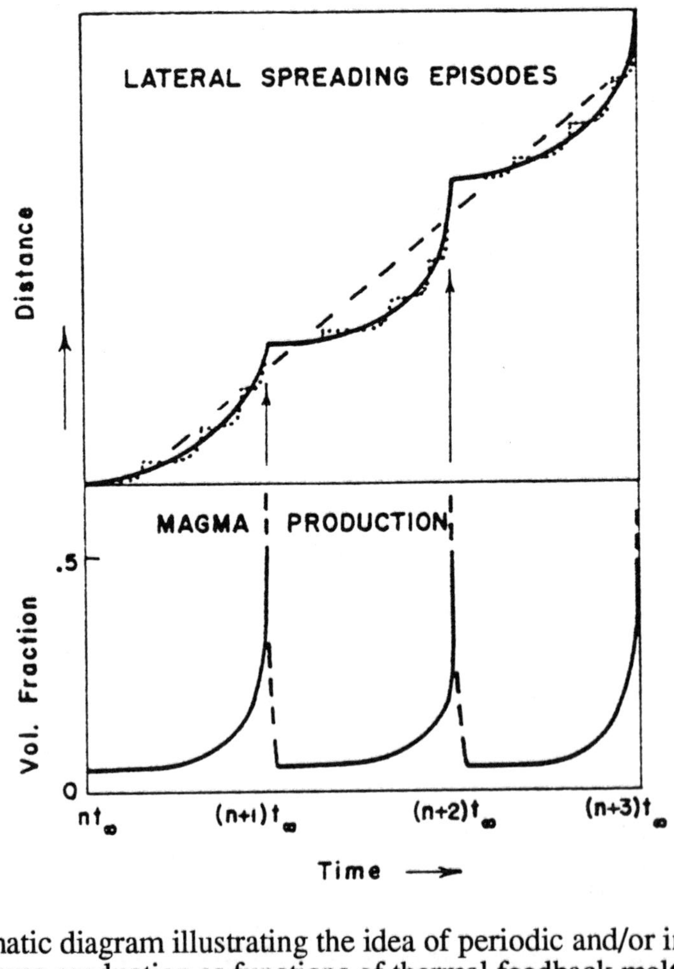

Figure 8.37: Schematic diagram illustrating the idea of periodic and/or intermittent pulses of magma production as functions of thermal feedback melting in the mantle, following the suggestion by Shaw (1969); applied to plutonic processes by Shaw et al. (1971, Fig. 5) and volcanic processes by Shaw (1973). The effect is directly analogous to Fig. 8.1 (see discussion at end of Section 8.2).

8.4 1

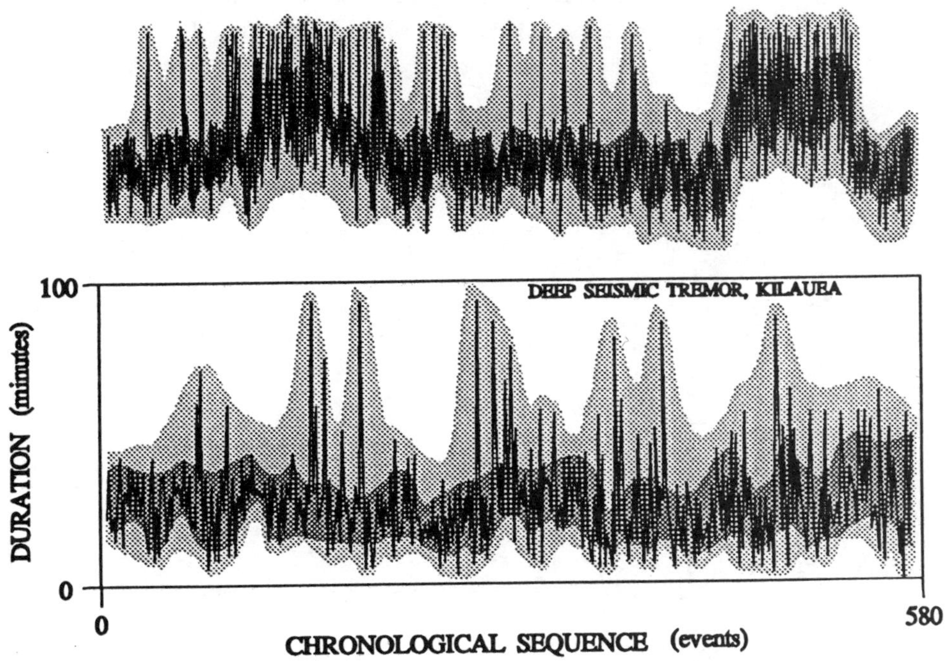

Figure 8.38: Intermittency of deep volcanic tremor in Hawaii, as analyzed by Shaw and Chouet (1988, 1989) who concluded that it has a multifractal signature consistent with the critical Golden Mean circle map analogous to the cases of Jensen et al. (1985) for coupled convection, and Gwinn and Westervelt (1987) for a coupled semiconductor device. The time scale of the tremor record is 22 years. Tremor intermittency is quantitatively explained by magmatic extensional-fracture percolation, suggested by Shaw and Chouet (1991) to have self-organized power-law scaling that extends to eruptive frequencies and even to long-term evolution of the volcanic chain. The unlabeled time series at the top is from Fig. 8.2, upside down (longer durations correlate with greater stress release).

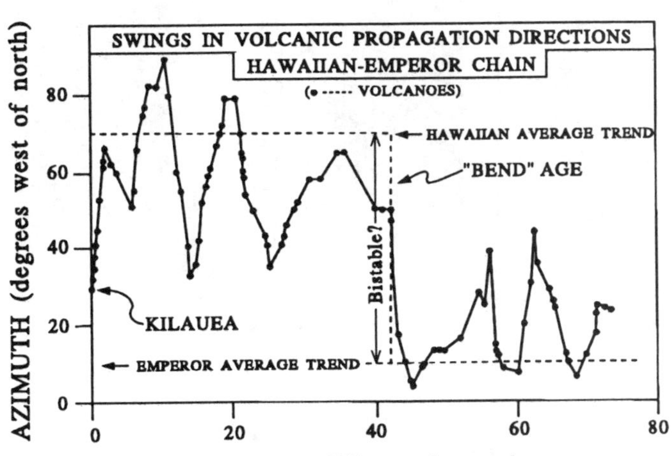

Figure 8.39: A possible case of bistability in Plate Tectonics? Plot of directional oscillations in the propagation of the Hawaiian-Emperor volcanic chain across the Pacific. Redrawn from Shaw et al. (1980).

8.42

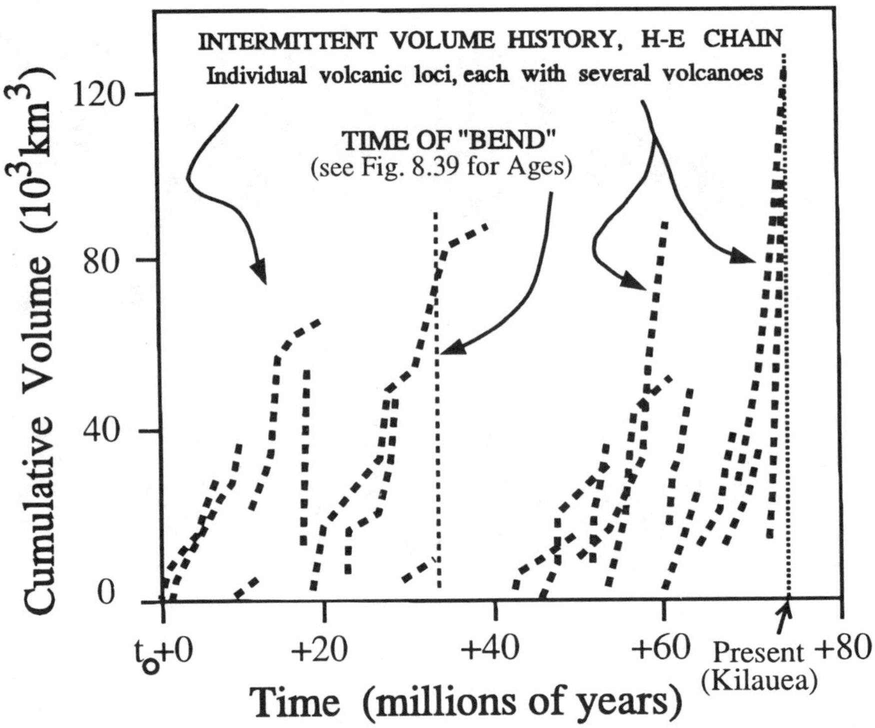

Figure 8.40: The intermittent evolution of volcanic loci along the Hawaiian-Emperor volcanic chain (modified from Shaw, 1987a, Figure 51.1). Note the repetitive nature of accelerative episodes, as in Fig. 8.37.

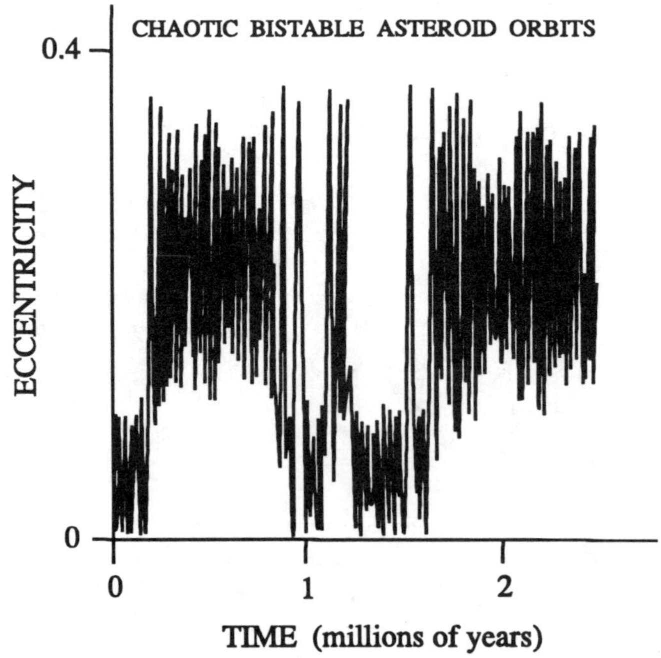

Figure 8.41: Chaotic bistability of asteroid orbital eccentricities, schematically redrawn from Wisdom (1987, Fig. 5).

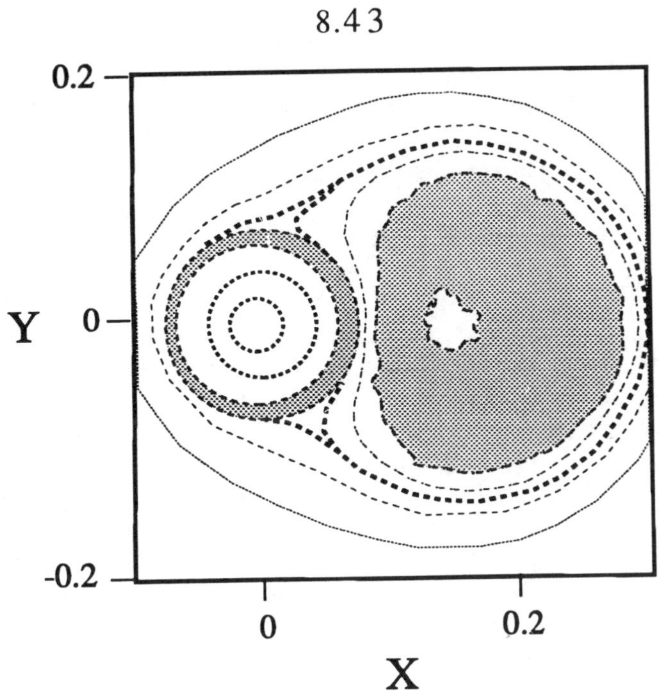

Figure 8.42: Caricature of phase portrait of chaotic asteroid orbital eccentricities after Wisdom (1987, Fig. 7), where the eccentricity is given by radial distance from the origin.

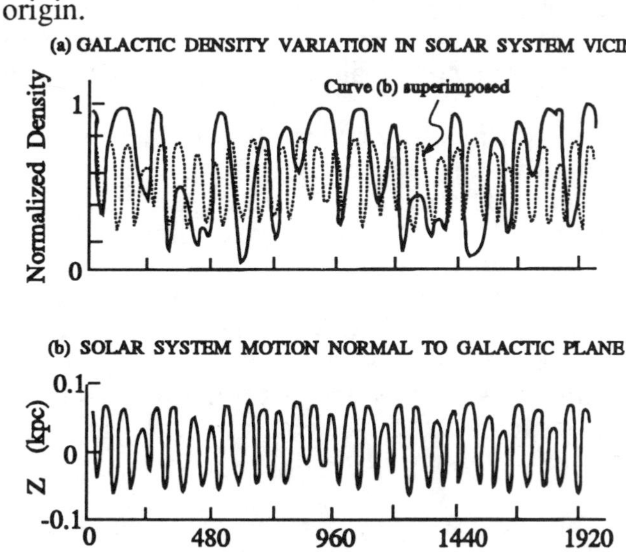

Figure 8.43: Calculated time series of galactic parameters in the vicinity of the Solar System over a time span of 2 billion years, schematically redrawn and modified after Innanen et al. (1978). Note resemblances to time series of earlier diagrams (especially, Figs. 8.7, 8.26, and 8.30) and implications of the superimposed curve as to possible nonlinear resonance phenomena in the Solar System. It is notable that the orbital parameters of Pluto show oscillations of orbital inclination with an average period of about 34 Ma (Wisdom, 1987, Fig. 14). This is roughly half the galactic-plane-crossing period and resembles the average period in many terrestrial phenomena (see Shaw, 1987b).

8.7 Note Added On Attractor Basins, Mixing, and Self-Organization: An Attractor Phase Portrait Is Not Just Another Pretty Picture.

It was noticed after the manuscript was formatted that the statement on page 8.9 which refers to the grey area in Figure 8.3 as the *attractor basin* is incomplete. The actual attractor basin also includes that portion of the surrounding fluid in which any initial position of a marker will be drawn into the circulation within the grey area, which is the limit of the *attractor* insofar as it was defined by the trajectories of markers as they were observed in the experiment. Between the grey area and the walls, rotor, and free surface any neutrally bouyant particles introduced into the fluid entered into the general circulation as shown. Near any of the interfaces, especially the outer walls and bottom, however, there is the possibility of attachment due to other attracting forces (interfacial energies). This might also be true of markers caught up with surface energy effects near the free surface and/or in the vicinities of cavitation. The latter would presumably represent topological bifurcations in phase space relative to the trajectories shown. Barring unusually strong normal forces near the rotor, it is a repelling locus because of the steep gradient of shear and viscous stresses near the center of rotation. Thus, the outer limit of the attractor basin would be some poorly defined, and probably fractal, surface between the grey area and the boundaries of the fluid. Lacking such artificial boundaries, the attractor basin would depend on the nature of other flow regimes in its vicinity. In their absence, it extends indefinitely, or to where other possible topological effects are generated by the induced motion. For example, if one stable and regenerative (non-decaying) vortical flow in an infinite homogeneous 'real' fluid is an attractor, then its basin of attraction must be very large. However, if other attractors exist, it seems necessary that individual particles may exchange diffusively with adjacent attractors, especially across fractal basin boundaries, so that the particles themselves do not remain for all time within one 'basin of attraction'.

Thus, actual basins of attraction for multicomponent flows must be leaky to an extent that depends on the spectra of kinetic energies of different types of particles. Attractor basins that have different properties for different types of particles presumably are composite or nested in some sense. Planetary atmospheres are examples of a global system of attraction in which there are many types of dissipative processes and possibilities for self-tuned stationary or time-dependent attractors and yet where individual particles are both gained and lost by the global system (i.e., there is a resemblance to 'living' systems). Expressed as a phase portrait of all particle motions, the global basin boundary would be an exceedingly complex multifractal interface that is intertwined throughout the Solar System with other planetary attractors, and we would have to include interstellar distances if cometary and related interactions are allowed.

Problems of mixing and of describing trajectories of motion of particles in different types of dynamical systems are both visually and mathematically difficult. In nonlinear dynamics they often are described by distinctions between: (a) *Hamiltonian systems* that are *integrable* as well as conservative, where "conservation" refers to preservation of area in phase space, (b) Hamiltonian systems that are *nonintegrable* even though they are area-preserving, and (c) *dissipative systems* (physically and/or in the phase space of numerical systems) that are "nonconservative" in the sense of not preserving area in phase space though energy may be conserved thermodynamically as in the thermosyphon loop, and are generally nonintegrable as in *noninvertible trajectories* of the sine circle map for $K > 1$. For instance, many problems of vortex structures and mixing are described kinematically and/or by Hamiltonian mechanics in both viscous and inviscid flows (e.g., Ottino, 1989, for mixing of viscous fluids; Marcus, 1990, for inviscid zonally 'shearing' flows). Marcus (1990) describes vortices analogous to Jupiter's "Great Red Spot" by integrating the equations of mass, momentum, and energy for an inviscid fluid. Is The Great Red Spot an attractor? Not formally, as described by these simulations, even though there are kinematic properties whereby smaller vortices can be drawn into and/or exchange angular momentum with larger more stable ones in a manner that could be called self-organization. This happens, however, according to prescribed sets of initial and boundary conditions. Such systems, therefore, though complex, are kindred to the area-preserving cases mentioned above. Marcus (1990, p. 415) discusses effects that can cause contraction of vortex area, showing that the net effect is small. The introductory remarks by Helleman (1983) are of interest with regard to distinctions between conservative and dissipative dynamical systems, and resemble those of this paper with respect to behavioral realities that may be neglected in University courses in Classical Mechanics.

Mathematically, viscous dissipation is recognized as present in modeling large scale fluid structures, but it is excluded to simplify calculations and/or to emphasize kinematic patterns that are assumed to be dominated by the mechanical equations of motion that are largely independent of viscous tractions and related thermal activations involving intermolecular diffusive motion and chemical reactions. If mechanisms of *dissipative self-organization* are of interest, however, involving the physics of thermomechanical feedback and autocatalytic processes, the possible evolution of *dissipative structures* is implicitly denied in such systems (cf., Prigogine and Stengers, 1984). Consequently, methods (a) and (b) categorically do not address the existence of *evolutionary systems* even though they may describe a progression of dynamical regimes to be anticipated from initial and/or boundary conditions and their changes. That is, *there are no attractors in such systems* in the sense of the convergence of initial trajectories from within larger basins of attraction or across complex fractal boundaries between regions of attraction in systems of multiple, possibly time-dependent, self-tuned attractors. From the viewpoint espoused in the introduction, *natural evolutionary systems* are those in which both the initial and boundary conditions are outcomes of the evolution of the nonlinear system of attraction itself.

All of the above systems can have complex periodic and/or chaotic regions, and the structural distinctions, for example, between area-preserving and dissipative systems are sometimes subtle and/or are not addressed (e.g., the resemblances of bistable orbits discussed earlier in viscous, mechanical, and planetary contexts). In the case of Hamiltonian systems, complex structures reflect local complexities of the kinematics demanded by the dynamical fields assumed to be acting on the system as a whole, but these are only partial descriptions of a fully coupled system of generalized Navier-Stokes equations (e.g., Bird et al., 1960). In dissipative systems, structural complexity represents both that imposed by the dynamical fields as well as that brought about by particle interactions in which irreversible thermal and other entropic effects are essential. Although the Hamiltonian idealization is a useful device for mapping partial regimes of many theoretical systems, it is an incomplete description from the standpoint of thermomechanical and rheological processes.

No real system is entirely devoid of thermomechanical dissipation, and here we begin to see the integrating power of irreversible thermodynamic effects toward the production of coherent structures. Some effects are analogous to laser systems in which highly pumped multimode energy sources are *slaved* to more coherent frequencies (cf., Haken, 1978; Abraham et al., 1985; Harrison and Biswas, 1986). It was suggested by Shaw (1988) that some aspects of ash-flow magmatism can be viewed in terms of *pumped systems far from equilibrium*. There a silicic focal magma system is driven by the power of multimode basaltic inputs to a coherent state of focused eruption at a variety of self-similar frequencies (cf., Smith, 1979). The experimental study of stable vortex growth in highly pumped turbulent rotating fluids by Sommeria et al. (1988, p. 690) designed to simulate Jupiter's Great Red Spot, the results of which resemble the numerical simulations by Marcus (1988) designed for the same purpose, also demonstrates a form of slaving of the turbulent energy sources to sequences of larger vortices with systematic sets of coherent spatial frequencies. The notion of slaving as a possibly general attribute of highly driven dynamical systems first occurred to me in simple numerical experiments with two coupled quadratic equations, where it was noticed that driving one or both equations toward larger tuning parameters (in one case holding one equation just above the critical transition to chaos) increased the order of the system by producing sets of *period-halving* attractors (e. g., Shaw, 1988, Figs. 9 and 10). Such experiments demonstrate how sensitive chaotic systems can be, under some circumstances, to small changes in the driving parameters that represent the dissipation, while under other circumstances some states of the system can be remarkably robust to large changes in the driving parameters.

The meaning of 'mixing' may be noted to have several different connotations. The kind of mixing of particle trajectories computed by Crutchfield et al. (1986) in the Lorenz system, referred to earlier, differs from that of fluid streaks that are simply stretched and folded but otherwise retain interfacial coherence and continuity (e.g., Ottino, 1989). In magmatic systems both types of effects can occur and may have greatly different implications as to the nature of chemical zoning in magma chambers, especially as the latter may affect the nature of an evolutionary path toward conditions of eruption (cf., Shaw, 1974; Smith, 1979; Huppert and Turner, 1981; Shaw, 1988). In order to see these effects clearly, 'particle' paths of diverse types must be demonstrated rather than inferred from such things as 'stream lines' or even from the limiting trajectories of a stable attractor or strange attractor (although the latter implies complex motions by virtue of its fractal structure, as mentioned in discussing the Lorenz strange attractor, Fig. 8.24). I did not emphasize this aspect sufficiently in Figures 8.3 through 8.5. Experiments using two marker particles (one red, one blue) showed that over some intervals of time while the blue marker was cycling outward, the red marker was cycling inward relative to the toroidal core region of Figure 8.3. This observation, together with the inward-outward spiral motion of Figure 8.4, the 'orbit' paths of Figure 8.5, and the range of rotational frequencies shown in that figure, demonstrate that the particle trajectories cross and interweave in intricate ways, inevitably producing a fractal geometric form that is almost impossible to envision without plotting it explicitly. Certainly, this interwoven form would not be anticipated from assumptions of either laminar or turbulent flow in such a highly viscous fluid, nor would it be if the 'mixing' were described kinematically or with a Hamiltonian formulation.

There is little surprise in seeing controversies in the literature that focus on issues of proof as to whether the histories of phenomena such as weather, climate, glacial, and volcanic cycles, etc., reflect the existences of 'low-dimensional attractors' in nature, when that 'nature' is evaluated against dynamical states that have been arbitrarily simulated in relatively few laboratory and/or computer experiments. It is far more interesting and informative to me that there are *fragments* of evidence in natural systems of vastly different material properties and scaling in time and space that attest to the same *types* of structural regimes that are seen in experimental and numerical nonlinear dynamics. This observation, when more fully documented, would lead to the almost inescapable conclusion that nature attests to the existence of an interrelated and variably self-tuned system of natural attractors (i.e., dissipative structures) that have in common the same sets of universal constants that have been discovered (and are yet to be discovered) in numerical dynamics (cf., Chouet and Shaw, 1991; Shaw and Chouet, 1991). The same natural system may simultaneously contain within it records of period-doubling, quasiperiodicity, mode-locking, and multiple chaotic periodicities and aperiodicities.

8.47

This situation would be analogous to a computer experiment with the sine circle map in which the tuning parameters K and ω, described earlier, each are cyclically driven across a unit range of values that spans the critical condition K = 1. Such an experiment may be particularly revelatory when it is realized that the natural analogs of the parameters K and ω also are analogous to the cyclic outputs of another circle map that determines their values, ad infinitum. *In such a concatenation, the properties that will be most conspicuous, rather than specific attractor states, are the meaures of self-organization as expressed by their arrays of universal constants.* In the hypothetical case of the catenated circle map, several universal constants are revealed as interactively important. These are: π, the Golden Mean (Shaw, 1987b), the 'Feigenbaum ratios' (Feigenbaum, 1980), the fractal invariances of mode-locked 'Devil's Staircases', period-doubling sets, critical quasi-periodic sets, and so on (see discussion in Shaw, 1987b).

Shaw and Chouet (1991) have presented evidence suggesting that critical self-organization of magma transport processes in Hawaiian volcanism has power-law scaling of frequency bundles (e.g., Fein et al., 1985) in which the mean values correspond to periods ranging from that of tremor (about 1 sec) to long-term evolution of the Hawaiian-Emperor chain (of order 30 m.y.; see Fig. 8.39). Analogous scaling would be expected for the coupled silicic systems, mentioned above, that are slaved to the basaltic systems. It is tempting to suggest that self-organization may explain frequency correlations among analogous processes throughout the Solar System. At the end of Section 8.2 the discussion of thermomechanical instabilities mentioned a similarity between the frequencies of pulsations in solar flare activity analyzed by Kurths and Herzel (1987) and the frequency structure of near-surface volcanic tremor analyzed by Chouet and Shaw (1991). This seemingly farfetched comparison may not be unreasonable in the light of critical self-organization and universality. It is strengthened by the fact that the correlation dimension of an attractor that satisfied the solar data (the 'tremor' of solar eruptions) was about 3.5, while the mean value for the cited example of tremor in terrestrial volcanism is about 3.8. Both are low-dimensional attractors with similar degrees of freedom in phase space.

The ability to distinguish between the motions of complex attractors and other forms of complex motions that involve elements of mechanical and/or statistical independence is, in my view, the central issue of natural history, cosmological, geological, and biological. The growing debate over the existence of The Great Attractor in the distribution of galaxies (e.g., Dressler, 1991) underscores the cosmological context. It seems necessary that however the mass distribution may be rationalized in terms of classical methods of celestial mechanics, any preferred distribution of galaxies, not to mention their internal structures, calls for a principle of dissipative self-organization that operates on the largest scales of motion much as it does on the smallest. For example, adding to the possible universality mentioned above relative to solar and terrestrial 'tremor', Shaw and Chouet (1991) have proposed a model of fractally clustered magma distributions that resembles a model previously proposed for fractal clustering of stars and galaxies. The universality suggested by these examples speaks profoundly for the same organizational principles at other scales where it is often assumed that interdependence among 'particle' trajectories is negligible. I would include in this context principles of organic evolution in which it has been stated that "evolution has found no way to introduce novelty other than to produce a wide diversity of inheritable patterns *in some essentially random fashion*, and then to expose those patterns to the test of competitive survival" (Burnet, 1976, p. 158; cf., Shaw, 1987b); the italics were added. Other observations by Burnet (1976, p. 160) concerning the organizational principles of complex biological systems are similar to those of the present discussion.

8.48

Nonlinear dynamics provides new tools for the comparative study of systems with and without dissipation. For the first time methods exist by which we can begin to develop a comparative taxonomy of complex systems where the 'strengths' and domains of influence of basins of attraction in numerical and analog systems can simulate some aspects of self-tuning in natural systems. Thus, we have a new ability to test for universal scale invariance among patterns that represent self-organized dissipative structures and those that exist because of the choices of 'initial' and/or 'final' conditions that are integrated relative to assumed laws of motion. This distinction appears to represent the nexus at a fundamental level between diametrically opposed arguments concerning the nature of evolution in cosmological to biological contexts. Therefore, nonlinear dynamics can provide the ability to arbitrate evolutionary models in a manner not previously recognized.

8.8 Acknowledgements

I thank my colleagues Peggy Bruggman, Mike Sawlan, and Bob Tilling for help with the graphics and formatting. In the process of digging into old records, I was reminded anew of the rewards of doing experiments the old-fashioned way ("19th-century experiments" some of my friendly critics called them). But for that, I would not have encountered Irv Gruntfest who, wherever he may be, I thank again for challenging me to enquire more deeply than before into what was happening before my eyes. In the same vein, I again thank Bob Smith and Dale Jackson for like challenges in volcanology, and Tom Wright for catalyzing, and collaborating in, the field measurements of magma rheology in Hawaii, ---- and Bill Glen for reminding me of history, and daring me to expand my field of vision. Pat Doherty has been unstinting in his help with problems of computing. Without his help the fun of seeing numerical dynamics create form 'out of whole cloth' (or 'transfinity') would only have been vicarious. And last-but-not-least, without my current collaboration with Bernard Chouet on studies of volcanic tremor the ubiquitous importance of coupled oscillations, self-organization, and universality, --- the themes of this review, together with thermomechanical dissipation, --- would have remained far more conjectural than they now appear to be.

8.9 References

Abraham, F. D., 1990, A visual introduction to dynamical systems theory for psychology. Santa Cruz, CA, Aerial Press, 287 p.

Abraham, N. B., L. A. Lugiato, and L. M. Narducci, feature eds., 1985, Instabilities in active optical media. Jour. of the Optical Society of America, B, Optical Physics, v. 2 (no. 1), 264 p.

Abraham, R. H., and C. D. Shaw, 1982-88, Dynamics - the geometry of behavior. Parts 1-4, VISMATH, v. 1-4, Santa Cruz, CA, Aerial Press, 678 p. (4-vol. set).

Bak, P, and C, Tang, 1989, Earthquakes as self-organized critical phenomena. Jour. of Geophysical Research, v. 94, p. 15635-15637.

Baker, D. N., A. J. Klimas, R. L. McPherron, and J. Büchner, 1990, The evolution from weak to strong geomagnetic activity: an interpretation in terms of deterministic chaos. Geophysical Research Letters, v. 17, p. 41-44.

8.49

Bird, R. B., W. E. Stewart, and E. N. Lightfoot, 1960, Transport phenomena. N.Y., John Wiley and Sons, 780 p.

Burnet, M., 1976, A homeostatic and self-monitoring immune system, *in* Burnett, F. M., Immunology: readings from Scientific American. San Francisco, W. H. Freeman and Company, p. 158-161.

Cassen, P. M., S. J. Peale, and R. T. Reynolds, 1982, Structure and thermal evolution of the Galilean satellites, *in* Morrison, D., ed., Satellites of Jupiter. Tucson, University of Arizona Press, p. 93-128.

Chouet, B., and H. R. Shaw, 1991, Fractal properties of tremor and gas-piston events observed at Kilauea Volcano, Hawaii. Jour. of Geophysical Research, *in press*.

Creveling, H. F., J. F. De Paz, J. Y. Baladi, and R. J. Schoenhals, 1975, Stability characteristics of a single-phase free convection loop. Jour. of Fluid Mechanics, v. 67, p. 65-84.

Crutchfield, J. P., J. D. Farmer, N. H. Packard, and R. S. Shaw, 1986, Chaos. Scientific American, v. 255, p. 46-57.

Ding, E. J., and P. C. Hemmer, 1988, Winding numbers for the supercritical sine circle map. Physica D, v. 32, p. 153-160.

Ditto, W. L., S. Rauseo, R. Cawley, C. Grebogi, G -H. Hsu, E. Kostelich, E. Ott, H. T. Savage, R. Segnan, M. L. Spano, and J. A. Yorke, 1989, Experimental observation of crisis-induced intemittency and its critical exponent. Physical Review Letters, v. 63, p. 923-926.

Dressler, A., 1991, The Great Attractor: do galaxies trace the large-scale mass distribution? Nature, v. 350, p. 391-397.

Feigenbaum, M. J., 1980, Universal behavior on nonlinear systems. Los Alamos Science, v. 1, p. 4-27.

Fein, A. P., M. S. Heutmaker, and J. P. Gollub, 1985, Scaling at the transition from quasiperiodicity to chaos. Physica Scripta, v. T9, p. 79-84.

Gruntfest, I. J., and H. R. Shaw, 1974, Scale effects in the study of Earth tides. Trans. of the Society of Rheology, v. 18, p. 287-297.

Gwinn, E. G., and R. M. Westervelt, 1987, Scaling structure of attractors at the transition from quasiperiodicity to chaos in electronic transport in Ge. Physical Review Letters, v. 59, p. 157-160.

Haken, H., 1978, Synergetics, an introduction. N. Y., Springer-Verlag, 355 p.

Halsey, T. C., M. H. Jensen, L. P. Kadanoff, I. Procaccia, and B. I. Shraiman, 1986, Fractal measures and their singularities: the characterization of strange sets. Physical Review A, v. 33, p. 1141-1151.

8.50

Harrison, R. G., and D. J. Biswa, 1986, Chaos in light. Nature, v. 321, p. 394-401.

Helleman, R. H. G., 1980, Self-generated chaotic behavior in nonlinear mechanics, *in* Cohen, E. G. D., ed., Fundamental Problems in Statistical Mechanics, v. 5, N.Y., North-Holland Publ. Co., p. 165-233.

Holden, A. V., and M. A. Muhamad, 1986, A graphical zoo of strange and peculiar attractors, *in* Holden, A. V., ed., Chaos. Princeton, NJ, Princeton University Press, p. 15-35.

Hones, E. W. Jr., 1979, Transient phenomena in the magnetotail and their relation to substorms. Space Science Reviews, v. 23, p. 393-410.

Huppert, H. E., and D. R. Moore, 1976, Nonlinear double-diffusive convection. Jour. of Fluid Mechanics, v. 78, p. 821-854.

Huppert, H. E., and J. S. Turner, 1981, Double-diffusive convection. Jour. of Fluid Mechanics, v. 106, p. 299-329.

Innanen, K. A., A. T. Patrick, and W. W. Duley, 1978, The interaction of the spiral density wave and the Sun's galactic orbit. Astrophysics and Space Science, v. 51, p. 511-515.

Jensen, M. H., L. P. Kadanoff, A. Libschaber, I. Procaccia, and J. Stavans, 1985, Global universality at the onset of chaos: results of a forced Rayleigh-Bénard experiment. Physical Review Letters, v. 55, p. 2798-2801.

Kepler, T. B., S. Datt, R. B. Meyer, and L. F. Abbott, 1990, Chaos in a neural network circuit. Physica D, v. 46, p. 449-457.

Knobloch, E., D. R. Moore, J. Toomre, and N. O. Weiss, 1986, Transitions to chaos in two-dimensional double-diffusive convection. Jour. of Fluid Mechanics, v. 166, p. 409-448.

Kurths, J., and H. Herzel, 1987, An attractor in a solar time series. Physica, D, v. 25, p. 165-172.

Lauterborn, W., and J. Holzfuss, 1991, Acoustic chaos. International Jour. of Bifurcation and Chaos, v. 1, p. 13-26.

Lighthill, M. J., 1953, Theoretical considerations on free convection in tubes. Quarterly Jour. of Mechanics and Applied Mathematics, v. 5, p. 398-439.

Lorenz, E. N., 1963, Deterministic non-periodic flow. Jour. of the Atmospheric Sciences, v. 20, p. 130-141.

Lorenz, E. N., 1964, The problem of deducing the climate from the governing equations. Tellus, v. 16, p. 1-11.

Lorenz, E. N., 1979, On the prevalence of aperiodicity in simple systems. Lecture Notes in Mathematics., v. 755, p. 53-75.

8.51

Mandelbrot, B. B., 1982, The fractal geometry of nature. N.Y., Freeman, 468 p.

Marcus, P. S., 1988, Numerical simulation of Jupiter's Great Red Spot. Nature, v. 331, p. 693-696.

Marcus, P. S., 1990, Vortex dynamics in a shearing zonal flow. Jour. of Fluid Dynamics, v. 215, p. 393-430.

May, R. M., 1976, Simple mathematical models with very complicated dynamics. Nature, v. 261, p. 459-467.

Miles, J., 1984, Resonant motion of a spherical pendulum. Physica D, v. 11, p. 309-323.

Moon, F. C., 1987, Chaotic vibrations. N.Y., John Wiley and Sons, 309 p.

Moon, F. C., and G -X. Li, 1985, The fractal dimension of the two-well potential strange attractor. Physica D, v. 17, p. 99-108.

Moore, D. R., J. Toomre, E. Knobloch, and N. O. Weiss, 1983, Period doubling and chaos in partial differential equations for thermosolutal convection. Nature, v. 303, p. 663-667.

Ottino, J. M., 1989, The kinematics of mixing: stretching, chaos, and transport. N. Y., Cambridge University Press, 364 p.

Prigogine, I., and I. Stengers, 1984, Order out of chaos. N. Y., Bantam Books, 349 p.

Ruelle, D. and F. Takens, 1971, On the nature of turbulence. Commun. math. Phys., v. 20, p. 167-192.

Sandfort, J. F., 1962, Heat engines. N.Y., Doubleday and Co. 292 p.

Shaw, H. R., 1965, Comments on viscosity, crystal settling, and convection in granitic magmas. American Jour. of Science, v. 263, p. 120-152.

Shaw, H. R., 1969, Rheology of basalt in the melting range. Jour. of Petrology, v. 10, p. 510-535.

Shaw, H. R., 1970, Earth tides, global heat flow, and tectonics. Science, v. 168, p. 1084-1087.

Shaw, H. R., 1973, Mantle convection and volcanic periodicity in the Pacific; evidence from Hawaii. Geological Society of America Bulletin, v. 84, p. 1505-1526.

Shaw, H. R., 1974, Diffusion of H_2O in granitic liquids: Part I. Experimental data; Part II. Mass transfer in magma chambers, *in* Hofmann, A. W., B. J. Giletti, H. S. Yoder, Jr., and R. A. Yund, eds., Geochemical transport and kinetics. Carnegie Institution of Washington, p. 139-170.

8.52

Shaw, H. R., 1987a, Uniqueness of volcanic systems, *in* Decker, R. W., T. L. Wright, and P. H. Stauffer, Volcanism in Hawaii. U. S. Geological Survey, Professional Paper 1350, p. 1357-1394.

Shaw, H. R., 1987b, The periodic structure of the natural record, and nonlinear dynamics. Eos, v. 68, p. 1651-1665.

Shaw, H. R., 1988, Mathematical attractor theory and plutonic-volcanic episodicity, *in* King, C -Y, and R. Scarpa, Modeling of volcanic processes. Wiesbaden, F. Vieweg & Sohn, p. 162-206.

Shaw, H. R., and B. Chouet, 1988, Application of nonlinear dynamics to the history of seismic tremor at Kilauea Volcano, Hawaii. U. S. Geological Survey, Open-file Report 88-539, 78 p.

Shaw, H. R., and B. Chouet, 1989, Singularity spectrum of intermittent seismic tremor at Kilauea Volcano, Hawaii. Geophysical Research Letters, v. 16, p. 195-198.

Shaw, H. R., and B. Chouet, 1991, Fractal hierarchies of magma transport in Hawaii and critical self-organization of tremor. Jour. of Geophysical Research, *in press*.

Shaw, H. R., T. L. Wright, D. L. Peck, and R. Okamura, 1968, The viscosity of basaltic magma: an analysis of field measurements in Makaopuhi lava lake, Hawaii. American Jour. of Science, v. 266, p. 225-264.

Shaw, H. R., R. W. Kistler, and J. F. Evernden, 1971, Sierra Nevada plutonic cycle: Part II. Tidal energy and a hypothesis for orogenic-epeirogenic periodicities. Geol. Soc. America Bull., v. 82, p. 869-896.

Shaw, H. R., E. D. Jackson, and K. E. Bargar, 1980, Volcanic periodicity along the Hawaiian-Emperor chain. American Jour. of Science, v. 280-A, p. 667-708.

Shaw, R. S., 1984, The dripping faucet as a model chaotic system. Santa Cruz, CA, Aerial Press, 113 p.

Singer, J., Y -Z. Wang, and H. H. Bau, 1991, Controlling a chaotic system. Physical Review Letters, v. 66, p. 1123-1125.

Smith, R. L., 1979, Ash-flow magmatism. Geological Soc. America, Special Paper 180, p. 5-26.

Sommeria, J., S. D. Meyers, and H. L. Swinney, 1988, Laboratory simulation of Jupiter's Great Red Spot. Nature, v. 331, p. 689-693.

Sparrow, C., 1982, The Lorenz equations: bifurcations, chaos, and strange attractors. Berlin, Springer-Verlag, 269 p.

Sreenivasan, K. R., and C. Meneveau, 1986, The fractal facets of turbulence. Jour. of Fluid Mechanics, v. 173, p. 357-386.

Tritton, D. J., 1988, Physical fluid dynamics. Oxford, Clarendon Press, 519 p.

8.53

Van Buskirk, R., and C. Jeffries, 1985, Observation of chaotic dynamics of coupled nonlinear oscillators. Physical Review A, v. 31, p. 467-492.

Welander, P., 1967, On the oscillatory instability of a differentially heated fluid loop. Jour. of Fluid Mechanics, v. 29, p. 17-30.

Wisdom, J., 1987, Chaotic behavior in the Solar System, *in*, Berry, M. V., I. C. Percival, and N. O. Weiss, Dynamical chaos. Princeton, NJ, Princeton University Press, p. 109-129.

Wones, D. R., and H. R. Shaw, 1975, Tidal dissipation: a possible heat source for mare basalt magmas. Lunar Science VI, Abstracts of Papers, 6th Lunar Science Conference, The Lunar Science Institute, Houston, TX, p. 878-880.

Wu, X., and Z. A. Schelly, 1989, The effects of surface tension and temperature on the nonlinear dynamics of the dripping faucet. Physica D, v. 40, p. 433-443.

CHAPTER 9

NONLINEAR DYNAMICS OF CRUST AND MANTLE

Donald L. Turcotte
Department of Geological Sciences
Cornell University
Ithaca, NY 14853

9.1 **Introduction**

In this chapter we consider applications of the dynamical systems approach to two problems in solid-earth dynamics. The first problem we address is the deformation of the brittle upper crust. This deformation occurs primarily on pre-existing faults and takes place during earthquakes. It is recognized that the frequency-size distribution of earthquakes is fractal; it is inferred that the frequency-size distribution of faults is also fractal. There are several important questions that must be addressed. One is the behavior of individual faults. It is generally hypothesized that velocity-weakening friction on faults is responsible for their stick-slip behavior. This behavior can be represented by slider-block models. We will show the conditions under which slider-block models behave chaotically and will show that they follow the Feigenbaum, period-doubling route to chaos.

The second problem we will consider is mantle convection. Solid-state thermal convection in the earth's mantle is responsible for plate tectonics, mountain building, seismicity, and volcanism. An important characteristic of mantle convection is that the Prandtl number is essentially infinite. We address the question of whether infinite Prandtl number thermal convection is chaotic.

9.2 Chaotic Tectonics

Earthquakes generally occur on preexisting faults. A simple model for the behavior of a fault is a slider block pulled by a spring as illustrated in Figure 9.1 (Burridge and Knopoff, 1967). The block interacts with the surface through friction, this friction prevents sliding of the block until a critical value of the pulling force is reached. The block sticks and the force in the spring increases until it equals the frictional resistance to sliding on the surface, then slip occurs. This is analogous to an earthquake on a fault. This is stick-slip behavior. The stored elastic strain in the spring is relieved, this is known as elastic rebound on a fault. The behavior of the spring-block system will now be studied quantitatively. A constant velocity driver moving at velocity v extends the spring with spring constant k until the pulling force ky equals the frictional static resisting force F_s. The static condition for the onset of sliding is

$$ky = F_s \tag{9.1}$$

Once sliding begins the equation of motion for the block is

$$m\frac{d^2y}{dt^2} + ky = F_d \tag{9.2}$$

where m is the mass of the block and F_d is the dynamic friction. The sliding is analogous to an earthquake and relieves the stress in the spring in analogy to elastic rebound. The further assumption is made that the loading velocity of the driver v is sufficiently slow so that we may assume that it is zero during the sliding of the block. This is reasonable since an earthquake only lasts a few tens of seconds whereas the interval between earthquakes on a fault is typically hundreds of years or more.

The static-dynamic friction law is the simplest that generates stick-slip behavior, a necessary and sufficient condition for stick-slip behavior is that the static friction exceeds the dynamic friction, $F_s > F_d$. A variety of velocity-weakening friction laws are in agreement with laboratory observations and generate stick-slip behavior. Dynamic instabilities associated with complicated friction laws are well known from single-block models (Byerlee, 1978; Dietrich, 1981; Ruina, 1983; Rice and Tse, 1986). Slider-block models have been used to simulate foreshocks, aftershocks, pre- and post-seismic slip, and earthquake statistics (Dieterich, 1972; Rundle and Jackson, 1977; Cohen, 1977; Cao and Aki, 1984, 1986). Gu et al. (1984) found some chaotically bounded oscillations, Nussbaum and Ruina (1987) used a two-block model with spatial symmetry and found periodic behavior. Huang and Turcotte (1990) studied the same system without spatial symmetry and obtained classic chaotic behavior. Carlson and Langer (1989) used many blocks and also obtained chaotic behavior.

We first consider the solution for the behavior of the single block illustrated in Figure 9.1. It is convenient to introduce the nondimensional variables

$$\phi = \frac{F_s}{F_d} \quad \tau = t\left(\frac{k}{m}\right)^{1/2}, \; Y = \frac{ky}{F_s} \tag{9.3}$$

In terms of these variables the sliding condition (9.1) becomes

$$Y = 1 \tag{9.4}$$

and the equation of motion (9.2) becomes

$$\frac{d^2Y}{d\tau^2} + Y = \frac{1}{\phi} \tag{9.5}$$

We assume sliding starts at $\tau = 0$ with $Y = 1$ (9.4) and $dY/d\tau = 0$. The applicable solution is

$$Y = \frac{1}{\phi} + \left(1 - \frac{1}{\phi}\right)\cos\tau \tag{9.6}$$

Sliding ends at $\tau = \pi$ when $dY / d\tau$ is again 0. When the velocity is zero the friction jumps to its static value preventing further sliding. The position of the block at the end of sliding is $Y = (2/\phi) - 1$ so that the slip during sliding is

$$\Delta Y = 2\left(\frac{1}{\phi} - 1\right) \tag{9.7}$$

The dependence of Y and $dY/d\tau$ on τ during sliding are given in Figure 9.2 for

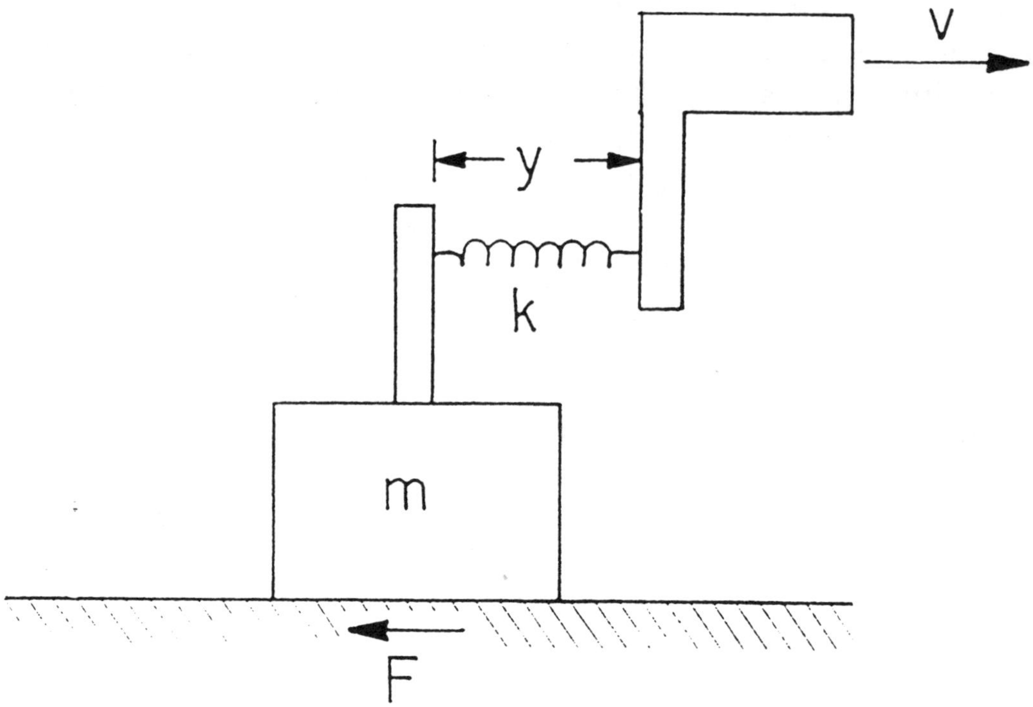

Figure 9.1 Illustration of the slider-block model for fault behavior. The constant velocity driver extends the spring until the force ky exceeds the static friction force F_S.

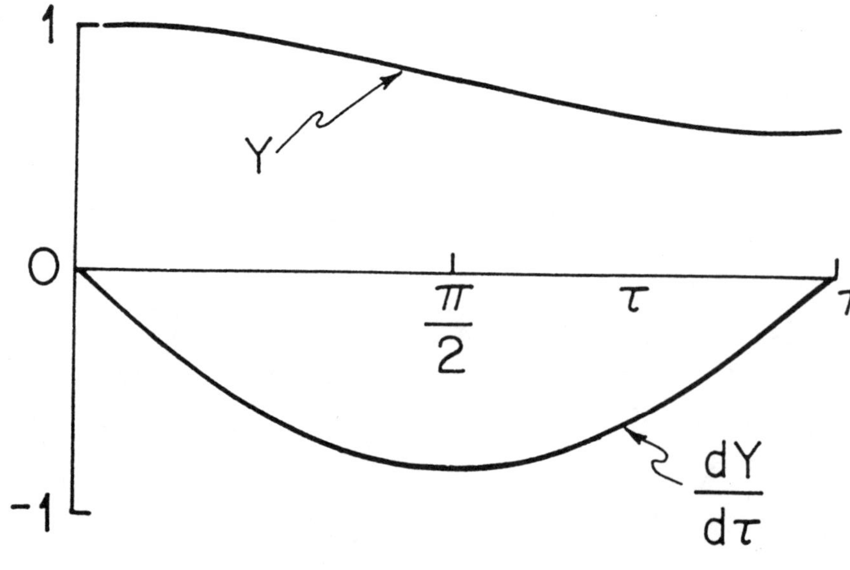

Figure 9.2 Dependence of Y and $dY/d\tau$ on τ during a sliding episode assuming $\phi = 1.25$.

$\phi = 1.25$. For this case Y drops from 1 to 0.6 during sliding and $\Delta Y = 0.4$. After sliding the spring extends due to the velocity of the driver until Y again equals 1 and the cycle repeats. With a single slider block periodic behavior is obtained. The variables Y and dY/dt define a phase plane for the solution.

We next consider the two-block system illustrated in Figure 9.3. Two sliding blocks are coupled to each other and to a constant velocity driver by linear elastic springs of stiffness k_c, k_1, and k_2, respectively. Other model parameters include the block masses m_1 and m_2, and the frictional forces F_1 and F_2. The position coordinates for the blocks, referred to the constant velocity driver, are y_1 and y_2. The two-block model is a fourth order system, y_1, y_2, v_1, v_2. It is the simplest system for which chaotic behavior might be expected.

The equations of motion for this two-block system are

$$m_1 \ddot{y}_1 + (k_1 + k_c)\, y_1 - k_c\, y_2 = F_1 \tag{9.8}$$

$$m_2 \ddot{y}_2 + (k_2 + k_c)\, y_2 - k_c\, y_1 = F_2 \tag{9.9}$$

Several additional assumptions are made to simplify the model. First, symmetry of the model is assumed except for the frictional forces. Accordingly, we take $m_1 = m_2 = m$, and $k_1 = k_2 = k$. In addition, the coupling spring constant is defined in terms of k: $k_c = \alpha k$. A second assumption is that the loading velocity of the driver is sufficiently slow that we may consider it to be zero during the sliding of the blocks. The ratio of the frictional forces of the two blocks β is introduced such that $F_2 = \beta F_1 = \beta F$. With these assumptions incorporated into the model, (9.8) and (9.9) can be written

$$m \ddot{y}_1 + (1 + \alpha)\, k y_1 - \alpha k y_2 = F \tag{9.10}$$

$$m \ddot{y}_2 + (1 + \alpha)\, k y_2 - \alpha k y_1 = \beta F \tag{9.11}$$

In this study, we choose a velocity-weakening friction law of the form

$$F = \frac{F_0}{1 + |\dot{y}/ v_f|} \tag{9.12}$$

The frictional forces decrease monotonically as velocity increases. In essence, this is a coupled oscillator system with damping from the velocity-weakening friction. Velocity-weakening phenomena have been observed in laboratory experiments (Tullis and Weeks, 1986).

It is convenient to introduce the nondimensional variables $Y_i = y_i k / F_0$ ($i = 1,2$), and a dimensionless time $\tau = t\sqrt{k/m}$. The failure criteria on Y_1 and Y_2 are now given by

$$Y_1 + \alpha (Y_1 - Y_2) = 1 \tag{9.13}$$

$$Y_2 + \alpha (Y_2 - Y_1) = \beta \tag{9.14}$$

and the equations of motion during slip are

$$\ddot{Y}_1 + Y_1 + \alpha (Y_1 - Y_2) = \frac{1}{1 + \gamma |\dot{Y}_1|} \tag{9.15}$$

$$\ddot{Y}_2 + Y_1 + \alpha (Y_2 - Y_1) = \frac{1}{1 + \gamma |\dot{Y}_2|} \tag{9.16}$$

where $\gamma = F_0 / v_f \sqrt{mk}$ is a measure of the amount of velocity weakening. These equations are numerically integrated using a fourth order Runge-Kutta scheme. The dynamic evolution of the system is recorded only after the system reaches a statistically steady state. Results illustrating the chaotic behavior of a spring-block system using static-dynamic friction have been given by Huang and Turcotte (1990).

We first consider the behavior of the system when it is symmetric, $\beta = 1$. After an initial transient the system exhibits cyclic behavior. Two examples are illustrated in Figure 9.4. The diagonal lines represent the failure envelopes for the two blocks given by (9.13) and (9.14) . The diagonal lines with unit slope represent the strain accumulation phases, the blocks stick and the springs extend at equal rates. Vertical lines represent the slip of block 2 (Y_1 constant) and horizontal lines represent the slip of block 1 (Y_2 constant). Slip occurs when the diagonal accumulation lines intersect the failure envelope. Slip continues until the block comes to rest, it then sticks. For the parameters chosen the blocks slip alternately in a cyclic manner. There are an infinite number of allowed solutions in which the slips are equal but the lengths of the strain accumulation phases vary. The particular limiting solution depends on the initial conditions.

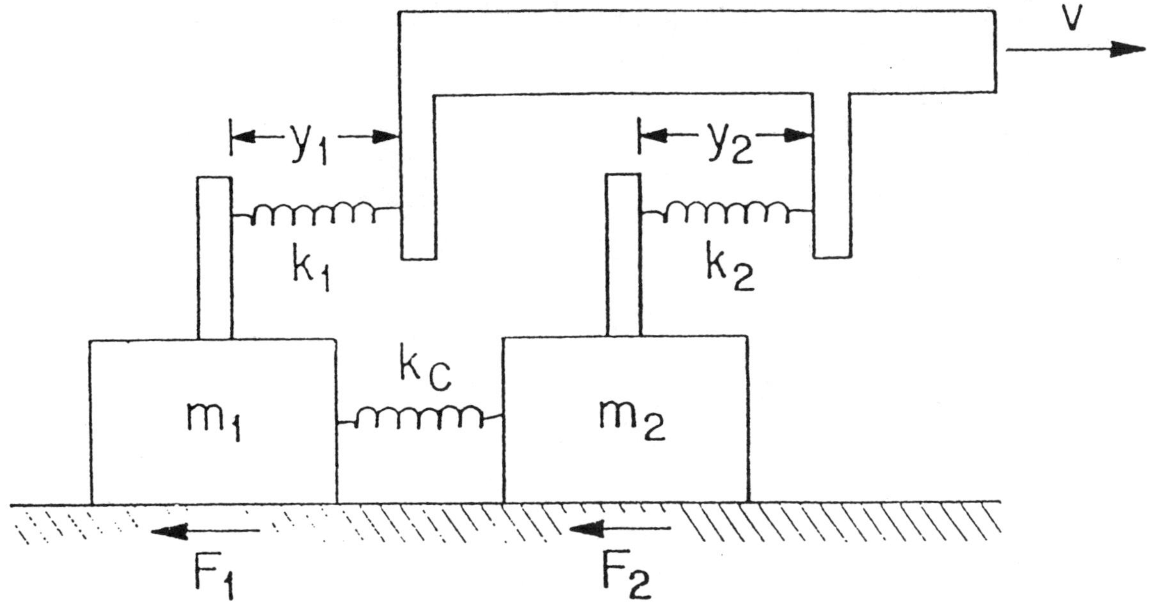

Figure 9.3 Illustration of the two-block model for fault interactions.

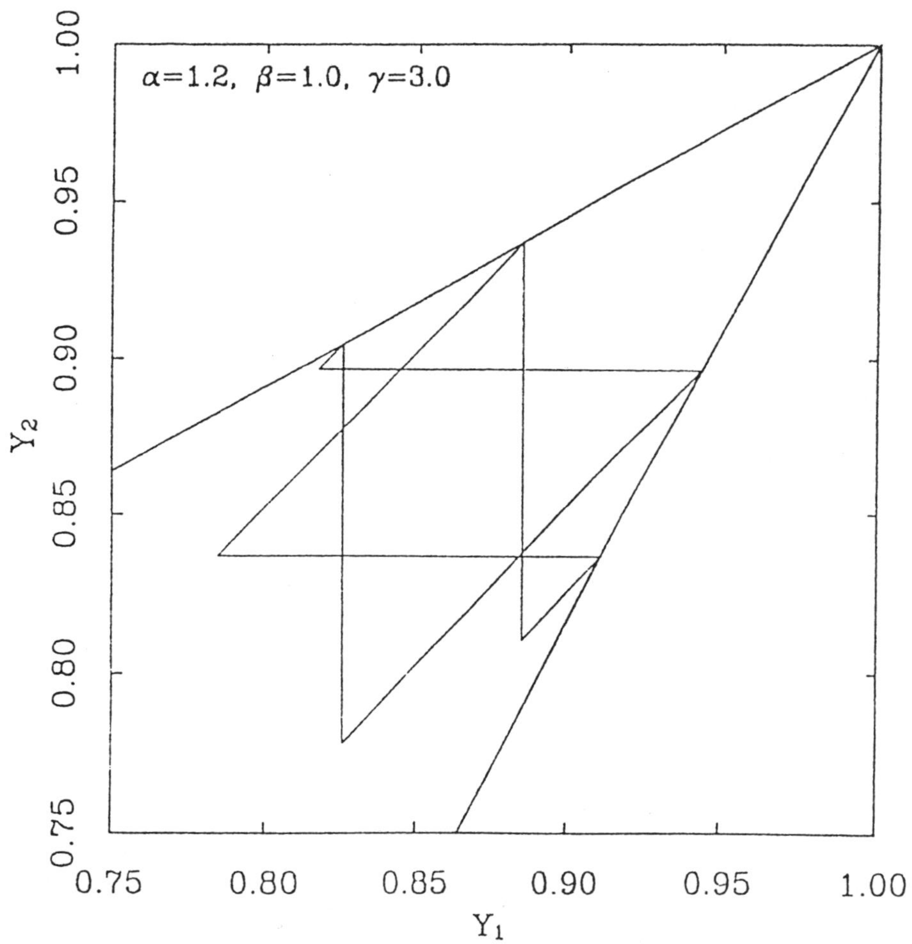

Figure 9.4 Phase portrait for a symmetrical two-block system. The phase plane coordinates are the position coordinates of the two blocks. Two examples are shown of the infinite number of allowed cycles. The particular cycle found depends upon the initial conditions for the run.

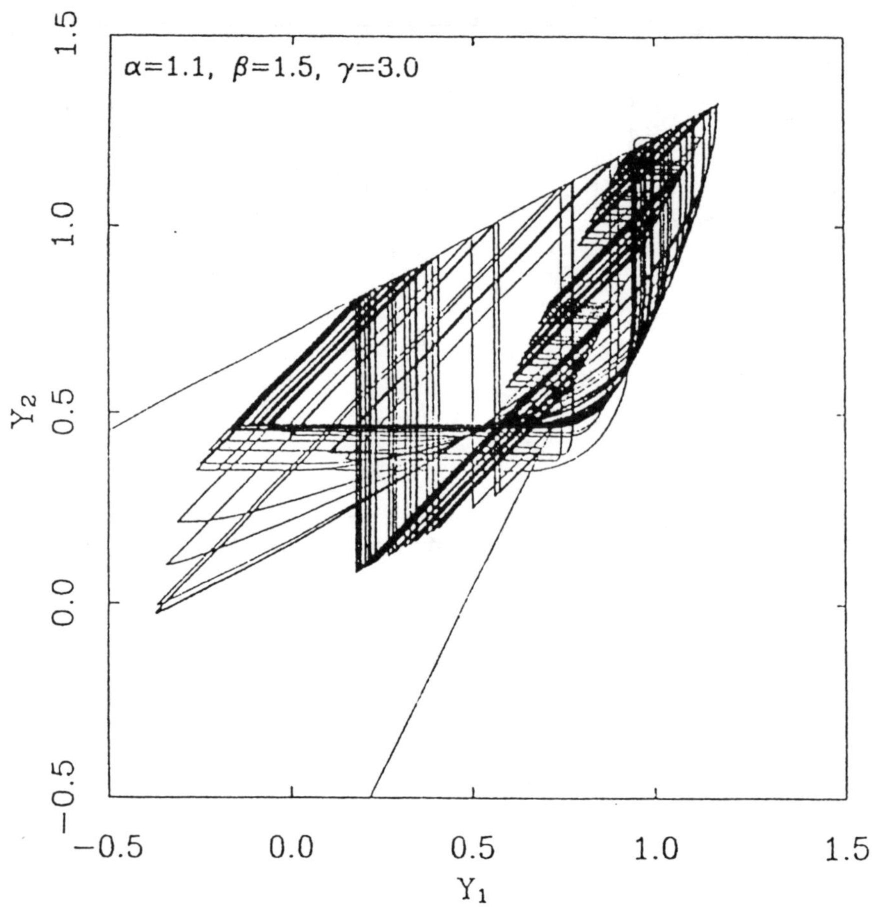

Figure 9.5 Phase portrait for an asymmetric two-block system. For these parameter values the evolution of the system is chaotic.

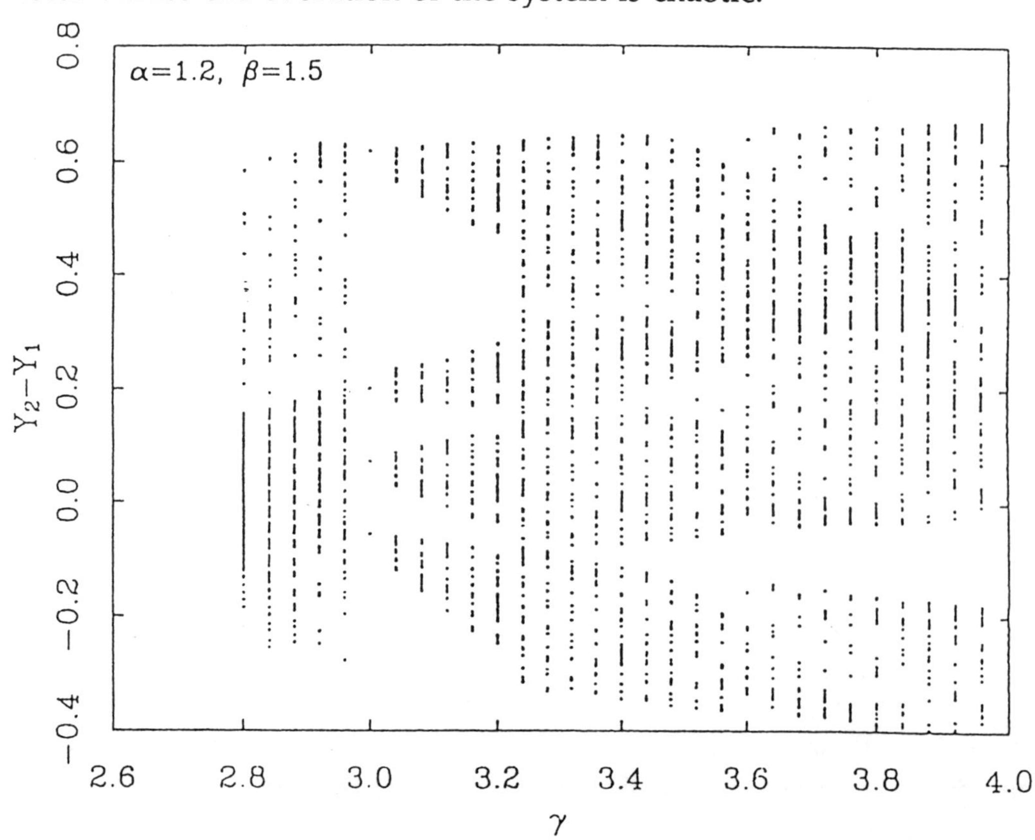

Figure 9.6 Bifurcation diagram giving the values of $Y_2 - Y_1$ after a large number of slip events for various values of the velocity-weakening parameter γ.

When the system is asymmetric, i.e. the ratio of frictional forces β is not equal to one, the dynamical picture changes completely. An example of a chaotic orbit with 100 events plotted in the phase plane (Y_1, Y_2) is shown in Figure 9.5. We note that there are single-block failures at some times, while at other times both blocks slide together. If an initial slip event from one failure line encounters the second failure line the second block also slips. Failure curves outside the failure envelope represent a multiple event.

In order to better illustrated the transition to chaos we consider the bifurcation diagram shown in Figure 9.6. For this example the velocity softening parameter γ is varied while holding the parameters α and β fixed. Similar results are obtained if the other parameters are varied. In Figure 9.6 the values $Y_2 - Y_1$ after a slip event are given for various values of γ. There appears to be a window of limit cycle behavior near $\gamma = 3$. Before examining this window in some detail we will consider the variation of the Lyapunov exponent under these conditions.

In a chaotic system, orbits that arise from nearby initial conditions diverge exponentially on the average. Lyapunov exponents measure the rates of divergence or convergence. Positive Lyapunov exponents indicate chaos while negative exponents indicate stability.

For a one-dimensional iteration map of known functional form $x_{n+1} = f'(x_n)$, the Lyapunov exponent λ is given by

$$\lambda = \lim_{n \to \infty} \frac{1}{n} \sum_{i=0}^{n-1} \log_2 |f'(x_i)| \tag{9.17}$$

A nonzero λ means that a pair of infinitesimally separated points initially a distance dx_0 apart diverge or converge exponentially on the average according to $dx_n = dx_0 2^{n\lambda}$. The derivation of (9.17) is straight forward if we notice the fact that $dx_{n+1} = f'(x_n)\, dx_n$, i.e. the derivative $f'(x_n)$ represents relative local stretching or contraction for each iteration. Then we have

$$2^{n\lambda} = \frac{dx_n}{dx_0} = \frac{dx_n}{dx_{n-1}} \cdots \frac{dx_1}{dx_0} = |f'(x_{n-1})|\,|f'(x_{n-2})| \cdots |f'(x_0)| \tag{9.18}$$

Values of the Lyapunov exponent λ corresponding to the bifurcation diagram, Figure 9.6, are given in Figure 9.7. The window of stable behavior near $\gamma = 3$ is clearly shown.

In order to examine the route to chaos in greater detail we illustrate a portion of the bifurcation diagram given in Figure 9.6 in the vicinity of $\gamma = 3$, this is done in Figure 9.8. This is a classic example of the period doubling route to chaos. Values of the Lyapunov exponent in this region are given in Figure 9.9. The phase portrait for the limit cycle at $\gamma = 3.000$ is given in Figure 9.10. Four slip events occur in this limit cycle. Two events involve the weaker block, one larger event involves the stronger block, and one large event involves both blocks. All initial conditions for the system evolve to this limit cycle for these parameter values.

At $\gamma = 3.00700$ the system experiences a flip bifurcation and the period doubles. This period doubling is clearly illustrated in Figure 9.8 and the Lyapunov exponent goes to zero at this value as expected. We find that we can determine the bifurcation point more accurately from the calculation of the Lyapunov exponent than from the bifurcation diagram. The phase portrait for the period-doubled limit cycle at $\gamma = 3.008$ is given in Figure 9.11.

In order to determine the further examples of period-doubling the detailed behavior of the Lyapunov exponent in the vicinity of $\gamma = 3.025$ is given in Figure 9.12. The subsequent examples of period doubling occur at $\gamma_2 = 3.0240$, $\gamma_3 = 3.02590$, and $\gamma_4 = 3.02654$. It is of interest to apply the Feigenbaum criteria for the period-doubling route to chaos for these results. This criteria is given by

$$\gamma_{ck} = \frac{F\gamma_{k+1} - \gamma_k}{F - 1} \tag{9.19}$$

where the Feigenbaum number $F = 4.669202$ and γ_c is the onset of chaotic behavior. Taking $\gamma_2 = 3.02240$ and $\gamma_1 = 3.00700$ we find $\gamma_{c1} = 3.2660$. Taking $\gamma_3 = 3.02590$ and $\gamma_2 = 3.02240$ we find $\gamma_{c2} = 3.02685$. Taking $\gamma_4 = 3.02654$ and $\gamma_3 = 3.02590$ we find $\gamma_{c3} = 3.02671$. Using the Feigenbaum criteria the predicted transition to chaos is in excellent agreement with the first positive Lyapunov exponent in Figure 9.12.

Slider block models can be constructed that exhibit self-organized criticality (Bak et al., 1988). According to this concept a natural system is in a marginally stable state, when perturbed from this state it will evolve

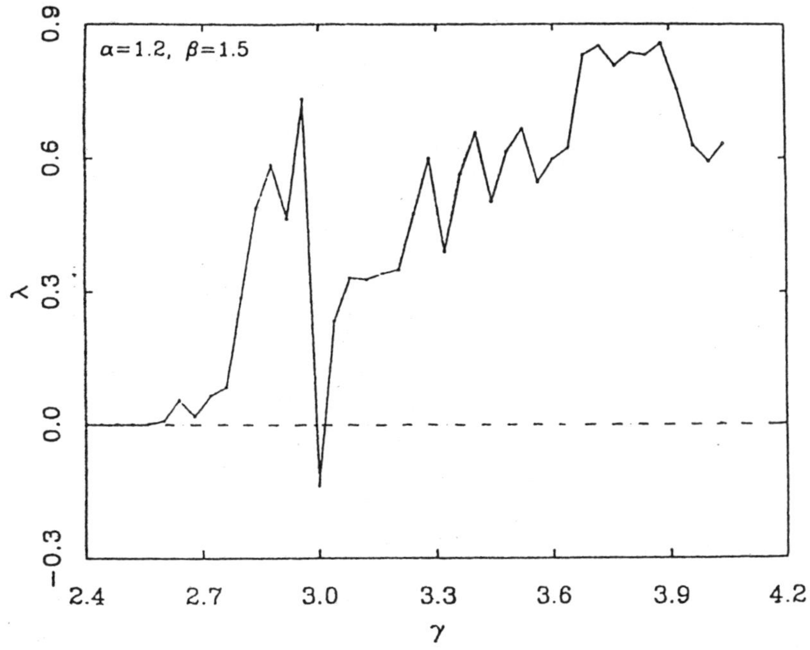

Figure 9.7 Dependence of the Lyapunov exponent λ on the velocity-weakening parameter γ for the same parameter range given in Figure 9.6.

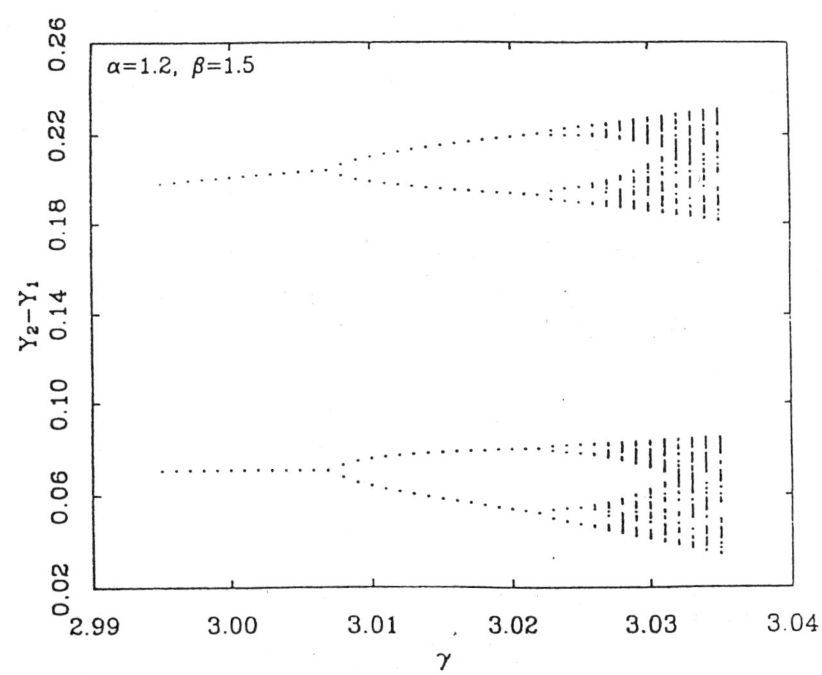

Figure 9.8 Details of the bifurcation diagram given in Figure 9.6 in the vicinity of the stable window at $\gamma = 3$.

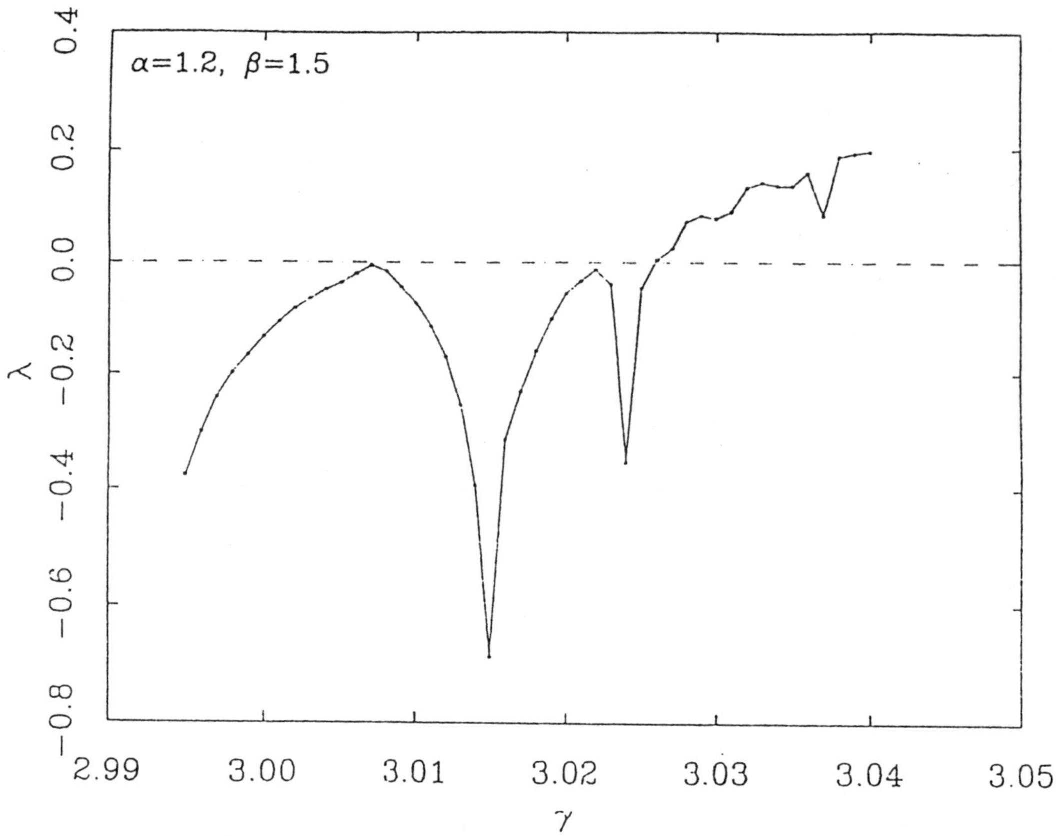

Figure 9.9 Values of the Lyapunov exponent from Figure 9.7 in the vicinity of the stable window at $\gamma = 3$.

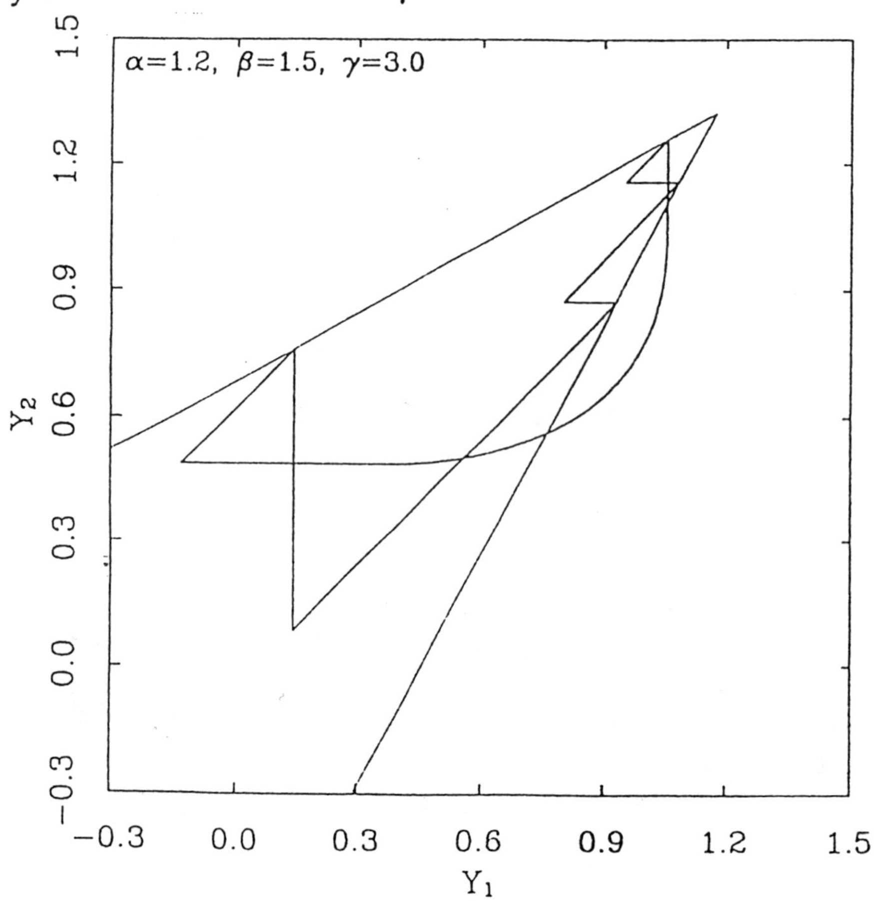

Figure 9.10 Phase portrait for the four event limit cycle at $\gamma = 3.000$.

naturally back to the state of marginal stability. In the critical state there is no longer a natural length scale so that fractal statistics are applicable. Also, the interaction length becomes large so that events can be correlated over large distances.

Above we showed that a pair of interacting slider blocks can exhibit deterministic chaos. This model is easily extended to include large numbers of slider blocks. (Carlson and Langer, 1989; Bak and Tang, 1989; Takayasu and Matsuzaki, 1988; Nakanishi, 1990; Ito and Matsuzaki, 1990; Sornette and Sornette, 1989, 1990). The slider blocks interact only with their nearest neighbors. When a block slips a fraction of the stored energy is lost due to friction and a fraction is transferred to the nearest neighbors. This transfer can lead to a multi-block failure. The frequency-size distribution of multiple block failures resemble those for cellular automata models and the Gutenberg-Richter relationship for earthquakes. It should be noted that there is one important difference between cellular automata models and the multiple slider-block models. The cellular automata models are statistical while the slider-block models can be completely deterministic.

The frequency-size distribution of events associated with self-organized criticality certainly resembles the regional distribution of earthquakes in a zone of active tectonics. This suggests that interactions between faults play an essential role in the behavior of such zones.

An important implication of a critical state in the crust is the range of interactions. A basic question is whether an earthquake on one part of planet, say Mexico, can trigger an earthquake at a large distance, say Japan. The classical approach to earthquakes would say that this is impossible. The stresses associated with seismic waves are too small to trigger an earthquake and there is no observational evidence for triggered events on this spatial scale. The stress changes associated with the fault displacement is localized and is damped by the athenospheric viscosity. However, interactions at large distances are a characteristic of critical phenomena. The interactions are not through the direct transmission of stress but through the interactions of faults with each other.

This action at a distance may help to explain the apparent success of the earthquake prediction algorithms developed at the International Institute for the Theory of Earthquake Prediction and Theoretical Geophysics in Moscow under the direction of Academician V.I. Keilis-Borok. This approach is based on pattern recognition of distributed regional seismicity

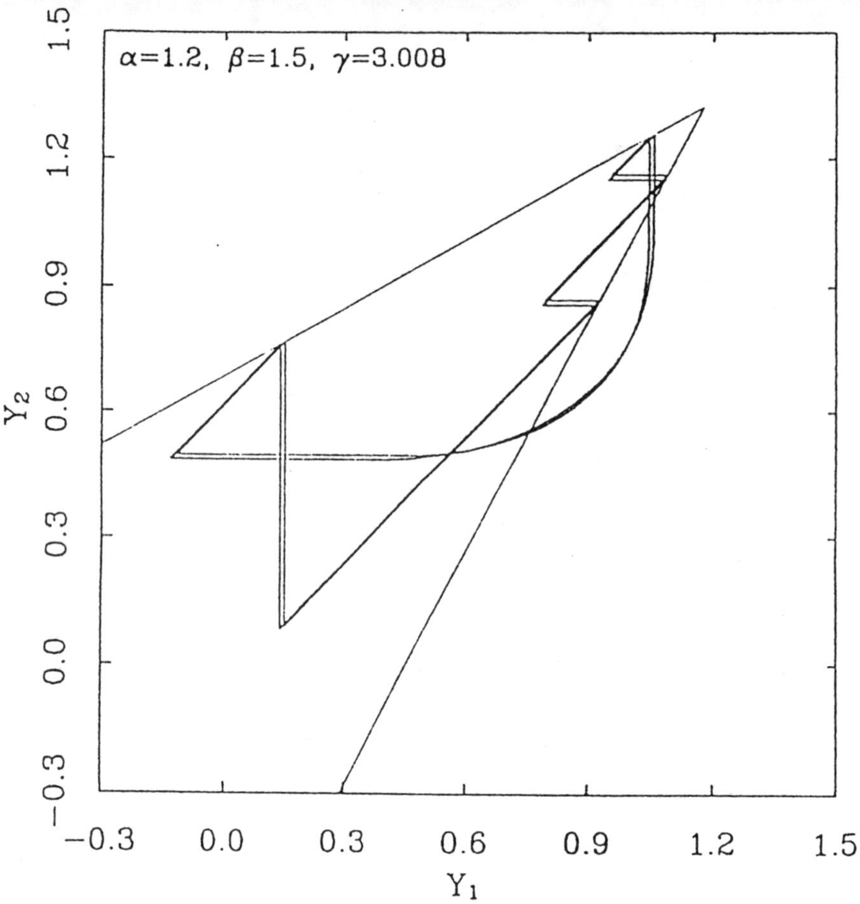

Figure 9.11 Phase portrait for the eight event period-doubled limit cycle at $\gamma = 3.008$.

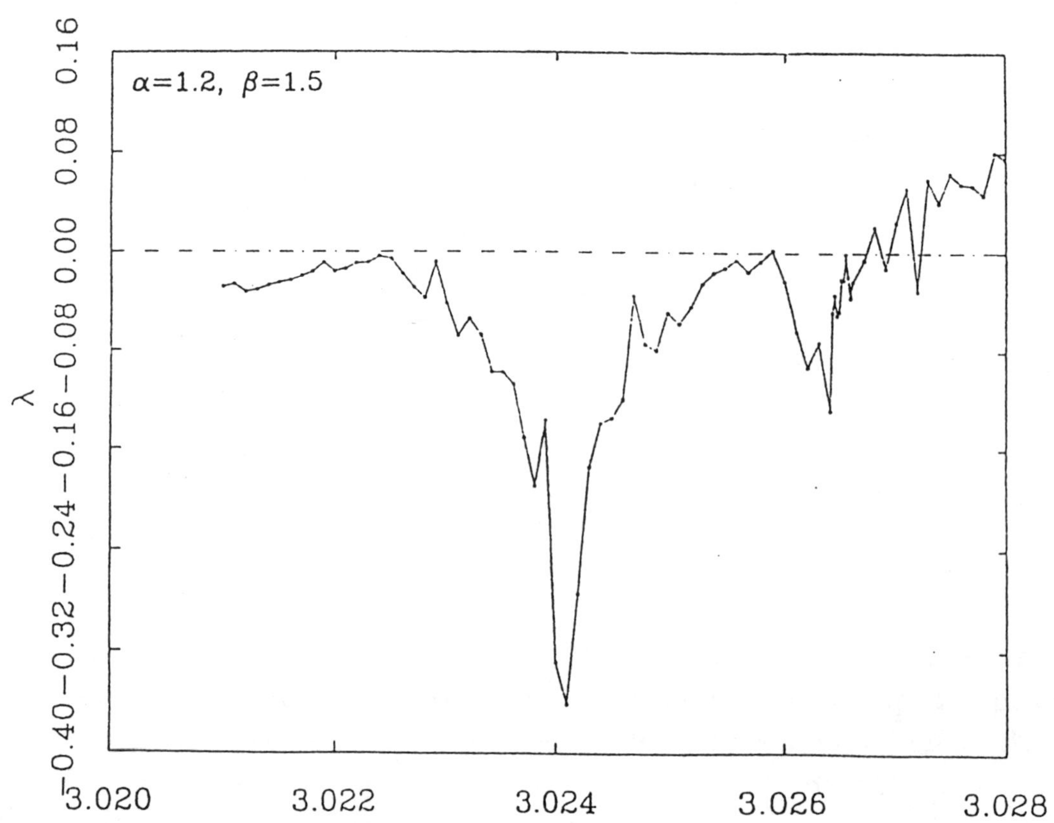

Figure 9.12 Values of the Lyapunov exponent from Figure 9.7 in the vicinity of the onset of chaotic behavior at $\gamma = 3.0267$.

(Keilis-Borok, 1990; Keilis-Borok and Rotwain, 1990; Keilis-Borok and Kossobokov, 1990). The pattern recognition includes quiescence (Schreider, 1990), increases in the clustering of events, and changes in aftershock statistics (Molchan et al., 1990). In reviewing regional seismicity after a major earthquake it is often observed that the regional seismicity in the vicinity of the fault rupture was anomalously low for several years prior to the earthquake. This is known as seismic quiescence. The problem has been to provide quantitative measures of quiescence prior to the major earthquake. The clustering of regional seismicity appears to become more fractal like prior to a large earthquake. There also appears to be a systematic reduction in the number of aftershocks associated with regional intermediate sized earthquakes prior to a major earthquake. Algorithms were developed to search earthquake catalogues for anomalous statistics over areas with diameters of 500 km. When a threshold of the anomalous behavior was reached a warning of the time of increased probability (TIP) of an earthquake was issued.

On a world-wide basis TIP's were triggered prior to 42 of 47 events. A TIP was released prior to both the Armenian earthquake on December 7, 1988 and the Loma Prieta earthquake on October 17, 1989. These are illustrated in Figure 12. A TIP was issued for region 3 in the Caucasus during January 1987 and was still in effect when the Armenian earthquake occurred in this region on December 7, 1988. A TIP was issued for region 5 in California during October 1984 and for region 6 during January 1985. These warnings were still in effect when the Loma Prieta earthquake occurred within both these overlapping regions on October 17, 1989.

The fault rupture of the Loma Prieta earthquake extended about 40 km. However, the prediction algorithms detected anomalous seismic behavior over two regions with diameters of 500 km. Self-organized criticality can explain anomalous correlated behavior over relatively large distances.

9.3 **Chaotic Mantle Convection**

Thermal convection in the earth's mantle transports the heat generated by the radioactive isotopes of uranium, thorium, and potassium and also cools the mantle. The surface plates of plate tectonics are the thermal boundary layers of mantle convection cells. The plates are created at ocean ridges, they cool and become gravitationally unstable with respect to the

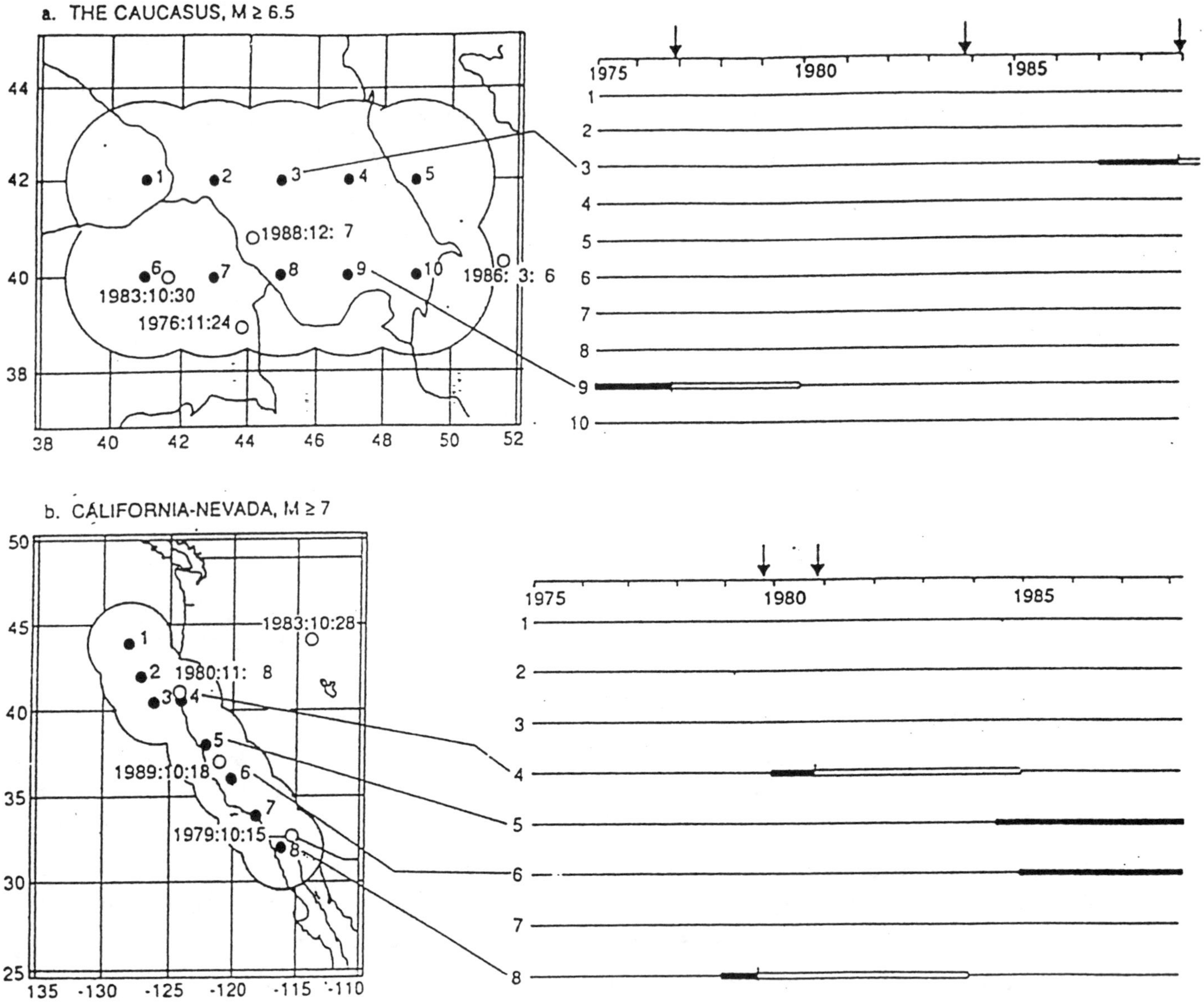

Figure 9.13 Illustrations of the Armenian (December 7, 1988) and Loma Prieta (October 18, 1989, Moscow time) earthquakes by Keilis-Borok (1990). In (a) the Caucasus region ia broken up in ten areas with diameters of 500 km; two warnings (for region 3 and 9) are shown on the right. The locations and times of four earthquakes are also given. In (b) the California-Nevada region is broken up into eight areas with diameters of 500 km. Four warnings (for regions 4-6, 8) and the locations and times of four earthquakes are given.

mantle below, and they founder and sink at ocean trenches. The plates transport about 80% of the heat associated with mantle convection but there is evidence for other mechanisms of mantle convection.

Intraplate hot spots such as Hawaii are attributed to hot plumes ascending from hot, unstable boundary layers at the base of the convecting mantle. The Hawaiian plume track extends to the Aleutian Islands indicating that this plume has been active for some 80 million years. This is an exception as most plume tracks are considerably shorter. Mantle convection is certainly unsteady, but is it "turbulent". In order to approach this question it is necessary to understand thermal convection at very large Prandtl numbers. The first approach is to consider a layer of Boussinesq fluid; this is a bad approximation for mantle convection for a number of reasons. The most important is that the viscosity of the mantle has an exponential dependence on temperature and pressure. In addition the mantle is compressible, the rheology may not be Newtonian, and a spherical shell should be considered. Nevertheless, the basic question concerning chaotic behavior can be addressed in the simple layered configuration with a Boussinesq fluid.

The equations governing thermal convection in a fluid layer of thickness h heated from below include conservation of mass

$$\frac{\partial u}{\partial x} + \frac{\partial v}{\partial y} = 0 \tag{9.20}$$

where u is the horizontal component of velocity in the x-direction and v is the vertical component of velocity in the y-direction, measured downward from the upper boundary. Conservation of momentum requires

$$\rho\left(\frac{\partial u}{\partial t} + u\frac{\partial u}{\partial x} + v\frac{\partial u}{\partial y}\right) = -\frac{\partial p}{\partial x} + \mu\left(\frac{\partial^2 u}{\partial y^2} + \frac{\partial^2 u}{\partial x^2}\right) \tag{9.21}$$

$$\rho\left(\frac{\partial v}{\partial t} + v\frac{\partial v}{\partial y} + u\frac{\partial v}{\partial x}\right) = -\frac{\partial p}{\partial y} + \mu\left(\frac{\partial^2 v}{\partial y^2} + \frac{\partial^2 v}{\partial x^2}\right) + \Delta\rho g \tag{9.22}$$

where ρ is density, $\Delta\rho$ the density difference, p pressure, μ viscosity, and g the acceleration of gravity. The density difference $\Delta\rho/\rho$ is assumed to be small so that it can be neglected except in the body force term $\Delta\rho g$ of the vertical momentum equation (9.22). This is known as the Boussinesq approximation. Conservation of energy requires

$$\rho\, c_p\left(\frac{\partial T}{\partial t}+u\frac{\partial T}{\partial x}+v\frac{\partial T}{\partial y}\right)=k\left(\frac{\partial^2 T}{\partial x^2}+\frac{\partial^2 T}{\partial y^2}\right) \qquad (9.23)$$

where T is temperature, k thermal conductivity, and c_p is the specific heat at constant pressure.

In the absence of convection, u=v=0, the temperature is a linear conduction profile given by

$$T_c = T_1 + (T_2 - T_1)\frac{y}{h} \qquad (9.24)$$

where T_1 is the constant temperature of the upper boundary (y = 0) and T_2 is the constant temperature of the lower boundary (y = h). The problem is simplified if we introduce a stream function ψ and temperature difference θ according to

$$u = -\frac{\partial \psi}{\partial y}, \qquad v = \frac{\partial \psi}{\partial x} \qquad (9.25)$$

$$\theta = T - T_c = T - T_1 - (T_2 - T_1)\frac{y}{h} \qquad (9.26)$$

The density difference $\Delta\rho$ can be related to the temperature difference by

$$\Delta\rho = -\rho\,\alpha\,\theta \qquad (9.27)$$

where α is the volumetric coefficient of thermal expansion.

Substitution of (9.25) to (9.27) into (9.20) to (9.23) gives

$$\rho\left[\left(\frac{\partial^2}{\partial x^2}+\frac{\partial^2}{\partial y^2}\right)\frac{\partial \psi}{\partial t}+\frac{\partial \psi}{\partial x}\left(\frac{\partial^2}{\partial x^2}+\frac{\partial^2}{\partial y^2}\right)\frac{\partial \psi}{\partial y}-\frac{\partial \psi}{\partial y}\left(\frac{\partial^2}{\partial x^2}+\frac{\partial^2}{\partial y^2}\right)\frac{\partial \psi}{\partial x}\right]$$

$$=\mu\left(\frac{\partial^4 \psi}{\partial x^4}+2\frac{\partial^4 \psi}{\partial x^2\,\partial y^2}+\frac{\partial^4 \psi}{\partial y^4}\right)-\rho\,\alpha\, g\frac{\partial \theta}{\partial x} \qquad (9.28)$$

$$\rho\, c_p\left[\frac{\partial\,\theta}{\partial\,t} - \frac{\partial\,\psi}{\partial\,y}\frac{\partial\,\theta}{\partial\,x} + \frac{\partial\,\psi}{\partial\,x}\frac{(T_2 - T_1)}{h} + \frac{\partial\,\psi}{\partial\,x}\frac{\partial\,\theta}{\partial\,y}\right] = k\left(\frac{\partial^2}{\partial\,x^2} + \frac{\partial^2}{\partial\,y^2}\right)\theta \qquad (9.29)$$

The problem has been reduced to the solution of two partial differential equations for ψ and θ.

In order to better understand the roles of various terms it is appropriate to introduce the nondimensional variables

$$\bar{t} = \frac{\kappa\, t}{h^2},\ \bar{x} = \frac{x}{h},\ \bar{y} = \frac{y}{h},\ \bar{\psi} = \frac{\psi}{\kappa},\ \bar{\theta} = \frac{\theta}{T_2 - T_1} \qquad (9.30)$$

where $\kappa = k/\rho\, c_p$ is the thermal diffusivity. With these nondimensional variables two nondimensional parameters govern the behavior of the equations.

$$Ra = \frac{\rho\,\alpha\, g\,(T_2 - T_1)\, h^3}{\kappa\,\mu} \qquad (9.31)$$

$$Pr = \frac{\mu}{\rho\,\kappa} \qquad (9.32)$$

where Ra is the Rayleigh number and Pr is the Prandtl number. Substitution of (9.30) to (12.13) into (9.32) and (9.29) gives

$$\frac{1}{Pr}\left[\left(\frac{\partial^2}{\partial\,\bar{x}^2} + \frac{\partial^2}{\partial\,\bar{y}^2}\right)\frac{\partial\,\bar{\psi}}{\partial\,\bar{t}} + \frac{\partial\,\bar{\psi}}{\partial\,\bar{x}}\left(\frac{\partial^2}{\partial\,\bar{x}^2} + \frac{\partial^2}{\partial\,\bar{y}^2}\right)\frac{\partial\bar{\psi}}{\partial\,\bar{y}} - \frac{\partial\,\bar{\psi}}{\partial\,\bar{y}}\left(\frac{\partial^2}{\partial\,\bar{x}^2} + \frac{\partial^2}{\partial\,\bar{y}^2}\right)\frac{\partial\,\bar{\psi}}{\partial\,\bar{x}}\right]$$

$$= \frac{\partial^4\,\bar{\psi}}{\partial\,\bar{x}^4} + 2\frac{\partial^4\,\psi}{\partial\,\bar{x}^2\,\partial\bar{y}^2} + \frac{\partial^4\,\bar{\psi}}{\partial\,\bar{y}^4} - Ra\frac{\partial\,\bar{\theta}}{\partial\,\bar{x}} \qquad (9.33)$$

$$\frac{\partial\,\bar{\theta}}{\partial\,\bar{t}} - \frac{\partial\,\bar{\psi}}{\partial\,\bar{y}}\frac{\partial\,\bar{\theta}}{\partial\,\bar{x}} + \frac{\partial\,\bar{\psi}}{\partial\,\bar{x}} + \frac{\partial\,\bar{\psi}}{\partial\,\bar{x}}\frac{\partial\,\bar{\theta}}{\partial\,\bar{y}} = \left(\frac{\partial^2}{\partial\,\bar{x}^2} + \frac{\partial^2}{\partial\,\bar{y}^2}\right)\bar{\theta} \qquad (9.34)$$

The solution is determined by the two parameters Ra and Pr and the boundary conditions.

Following Saltzman (1962), the stream function and the temperature are expanded in double Fourier series in x and y with coefficients that are functions of time only:

$$\psi(x,y,t) = \sum_{m=-\infty}^{\infty} \sum_{n=-\infty}^{\infty} \psi_{m.n}(t) \exp\left[2\pi i\left(\frac{mx}{\lambda} + \frac{ny}{2}\right)\right] \tag{9.35}$$

$$\theta(x,y,t) = \sum_{m=-\infty}^{\infty} \sum_{n=-\infty}^{\infty} \theta_{m.n}(t) \exp\left[2\pi i\left(\frac{mx}{\lambda} + \frac{ny}{2}\right)\right] \tag{9.36}$$

where λ is the nondimensional wavelength of the fundamental mode. Lorenz (1963) truncated these equations to a single term in the stream-function expansion and a pair of terms in the temperature expansion. The Lorenz expansion is

$$\psi = \frac{1}{2^{1/2}\lambda}\left(4 + \lambda^2\right) A(\tau) \sin\left(\frac{2\pi x}{\lambda}\right) \sin \pi y \tag{9.37}$$

$$\theta = \frac{1}{\pi r}\left[C(\tau) \sin 2\pi y - 2^{1/2} B(\tau) \cos\left(\frac{2\pi x}{\lambda}\right) \sin \pi y\right] \tag{9.38}$$

where

$$\tau = \pi^2\left[1 + \left(\frac{2}{\lambda}\right)^2\right] t \tag{9.39}$$

$$r = \frac{Ra}{Ra_c} \tag{9.40}$$

where Ra_c is the critical Rayleigh number for the onset of convection. These equations satisfy the boundary conditions $\psi = \nabla^2\theta = \theta = 0$ at $y = 0,1$. In the limit of infinite Prandtl number the Lorenz equations become

$$A = B \tag{9.41}$$

$$\frac{dB}{d\tau} = (r - 1)B - BC \tag{9.42}$$

$$\frac{dC}{d\tau} = bC + B^2 \tag{9.43}$$

where $b = 4/\left(1 + [2/\lambda]^2\right)$. The steady state solution A = B = C corresponds to heat conduction without flow. An additional pair of allowed solutions is

$$A = B = \pm[b(r - 1)]^{1/2} \tag{9.44}$$

$$C = r - 1 \tag{9.45}$$

These solutions correspond to cellular rolls either rotating clockwise or counterclockwise. A stability analysis for the conduction solution shows that it is stable for $r < 1$ and unstable for $r > 1$, a pitchfork bifurcation occurs at $r = 1$ ($Ra = Ra_c = 27\,\pi^2/4$). The steady state solution consisting of cellular rolls is stable for the entire range $r > 1$. In the limit of infinite Prandtl number the Lorenz equations do not yield chaotic solutions. This is expected since in this limit they reduce to a pair of equations and two degree of freedom systems generally do not yield chaotic solutions.

In order to consider this problem further Stewart and Turcotte (1989) considered a higher order expansion of the Lorenz equations. In the limit of infinite Prandtl number the Fourier coefficients of the stream function $\psi_{m,n}$ can be related to the Fourier coefficients of the temperature $\theta_{m,n}$ using (9.33) with the result

$$\psi_{m,n} = \left(\frac{2\pi i m}{\lambda}\right)\frac{\theta_{m,n}}{\left(\frac{4\pi^2 m^2}{\lambda^2} + \pi^2 n^2\right)^2} \tag{9.46}$$

This result is then substituted into (9.34). The lowest consistent order of truncation beyond that used by Lorenz is m = 2 for the expansion in x (m = 0, 1, 2) and n = 4 for the expansion in y (n = 1, 2, 3, 4). This truncation yields a set of 12 ordinary differential equations for the time dependence of the temperature coefficients

$$\frac{d\,\theta_{m,n}}{dt} = -\sum_{p=-2}^{2}\sum_{q=-4}^{4}\frac{2\pi^2}{\lambda}(mq - np)\left(\frac{2\pi p}{\lambda}\right)\cdot\frac{\theta_{p,q}\theta_{m-p,\,n-q}}{\left[\left(4\pi^2 p^2/\lambda^2\right) + \pi^2 q^2\right]^2}$$

$$+\mathrm{Ra}\left(\frac{2\pi\, m}{\lambda}\right)^2 \cdot \frac{\theta_{m,n}}{\left[\left(4\pi^2 m^2/\lambda^2\right)+\pi^2 n^2\right]^2} - \pi^2\left(\frac{4m^2}{\lambda^2}+n^2\right)\theta_{m,n} \tag{9.47}$$

It is necessary to take twice the resolution in the vertical direction as in the horizontal direction in order to resolve the $u\partial\theta/\partial x$ term in the energy equation.

The time evolution of the twelve coefficients $\theta_{0,1}$, $\theta_{0,2}$, $\theta_{0,3}$, $\theta_{0,4}$, $\theta_{1,1}$, $\theta_{1,2}$, $\theta_{1,3}$, $\theta_{1,4}$, $\theta_{2,1}$, $\theta_{2,2}$, $\theta_{2,3}$, $\theta_{2,4}$ is found by integrating the twelve equations given by (9.47) numerically. The time evolution can be thought of as trajectories in a twelve dimensional phase space. It is convenient to project the 12-dimensional trajectories onto the two-dimensional phase-space consisting of $\theta_{1,1}$ and $\theta_{2,1}$. These correspond to the fundamental mode and the first subharmonic. There are two parameters in this problem, the Rayleigh number, Ra or r, and the wavelength λ. In this discussion solutions are given only for the critical value of the wavelength $\lambda = 2^{3/2}$.

At the subcritical Rayleigh numbers $0 < r < 1$ ($0 < \mathrm{Ra} < 657.512$) the only fixed point of the solution is at the origin and it is stable, there is no flow. For higher Rayleigh numbers, the two fixed points corresponding to clockwise and counterclockwise rotations in the fundamental model $\theta_{1,1}$ become stable. The steady state solution for $\mathrm{Ra} = 10^4$ ($r = 15.21$) is given in Table 9.1. It is seen that only six of the 12 coefficients are nonzero: $\theta_{0,2}$, $\theta_{0,4}$, $\theta_{1,1}$, $\theta_{1,3}$, $\theta_{2,2}$, and $\theta_{2,4}$. This solution was obtained by specifying an initial condition near the origin and studying the time evolution of the 12 coefficients using (9.47). This time evolution projected onto the $\theta_{1,1}$ - $\theta_{2,1}$ plane is given in Figure 9.14a. Although the subharmonic coefficient $\theta_{2,1}$ is zero at the fixed point, it is nonzero during the time evolution.

The steady-state solution for $\mathrm{Ra} = 3 \times 10^4$ ($r = 45.62$) is also given in Table 9.1. It is seen that only four of the twelve coefficients are nonzero: $\theta_{0,2}$, $\theta_{0,4}$, $\theta_{2,1}$, $\theta_{2,3}$. At this Rayleigh number the fundamental mode and its associated coefficients $\theta_{1,1}$, $\theta_{1,3}$, $\theta_{2,2}$ and $\theta_{2,4}$ are zero at the stable fixed point. The time evolution of the solution projected onto the $\theta_{1,1}$ - $\theta_{2,1}$ plane is given in Figure 9.14b. Finally, the steady-state solution for $\mathrm{Ra} = 4.3 \times 10^4$ ($r = 65.39$) is given in Figure 9.14c. It is seen that eight of the twelve coefficients are now nonzero: $\theta_{0,1}$, $\theta_{0,2}$, $\theta_{0,3}$, $\theta_{0,4}$, $\theta_{2,1}$, $\theta_{2,2}$, $\theta_{2,3}$, $\theta_{2,4}$. All of the $\theta_{1,n}$ coefficients are zero including the fundamental mode $\theta_{1,1}$. The time evolution of the solution projected onto the $\theta_{2,1}$ - $\theta_{1,1}$ plane is given in Figure 9.14c. It is seen that the evolution prior to entering the stable fixed point is much more complex; the solution oscillates in the positive $\theta_{2,1}$ quadrants before entering the negative quadrants.

The time evolution of the solution for Ra = 4.5 x 10^4 (r = 68.44) is given in Figure 9.14d; it is fully chaotic and no fixed points are stable. The flow alternates between aperiodic oscillations about the two fundamental modes (clockwise and counterclockwise) and the two subharmonic modes (clockwise and counterclockwise. All 12 coefficients are nonzero and are time-dependent. The time dependence of the $\theta_{2,1}$ and $\theta_{1,1}$ coefficients are shown in Figure 9.15. The resemblance to the time behavior of the Lorenz attractor is striking. Oscillatory behavior of the $\theta_{2,1}$ mode amplifies until the flow undergoes a burst in the fundamental $\theta_{1,1}$ mode, where it is briefly trapped before flipping into the $\theta_{2,1}$ mode with the opposite sense of rotation.

In order to better understand the transitions in the behavior of the time-dependent solutions, we calculated the location of the fixed points as a function the Rayleigh number. Figure 9.16 gives two projections of the loci of fixed points as a function of Rayleigh number of this system. The solid lines denote stable fixed points, the dashed lines unstable fixed points, and the open circles Hopf bifurcations. Figure 3a shows a projection on to the Ra, $\theta_{1,1}$ plane. The original system of equations is symmetric under the transformations $\theta_{1,1} \rightarrow - \theta_{1,1}$ and $\theta_{2,1} \rightarrow - \theta_{2,1}$; therefore we only show the positive quadrant.

The origin representing the conduction solution becomes unstable and bifurcates at Rayleigh number Ra = 657.512, spawning two stable symmetric solutions which contain no component in the $\theta_{2,1}$ mode. One branch of this solution appears in the positive quadrant and is labelled "$\theta_{1,1}$ pure" in Figure 9.16a to distinguish it from the mixed-mode solution where there is some contribution from the $\theta_{2,1}$ mode. Each branch becomes unstable and undergoes a subcritical pitchfork bifurcation at Ra = 3.802×10^4, producing four unstable mixed mode solutions, labeled "$\theta_{1,1}$ mixed". Each $\theta_{1,1}$ mixed-mode branch sweeps back to a saddle bifurcation at Ra = 1.909×10^4. This type of bifurcation configuration

	Ra		
	10^4	3×10^4	4.3×10^4
$r = Ra/R_c$	15.21	45.62	65.39
$\theta_{0,1}$	0.000	0.000	-2120.222
$\theta_{0,2}$	-1506.097	-4571.719	-6663.231
$\theta_{0,3}$	0.000	0.000	-366.705
$\theta_{0,4}$	-286.133	-2191.940	-3355.777
$\theta_{1,1}$	506.978	0.000	0.000
$\theta_{1,2}$	0.000	0.000	0.000
$\theta_{1,3}$	66.170	0.000	0.000
$\theta_{1,4}$	0.000	0.000	0.000
$\theta_{2,1}$	0.000	1117.832	-1415.752
$\theta_{2,2}$	426.445	0.000	-398.378
$\theta_{2,3}$	0.000	509.313	-623.561
$\theta_{2,4}$	105.531	0.000	-490.614

Table 9.1. Numerical values of Fourier coefficients of the fixed points of the 12-mode equations (9.47)

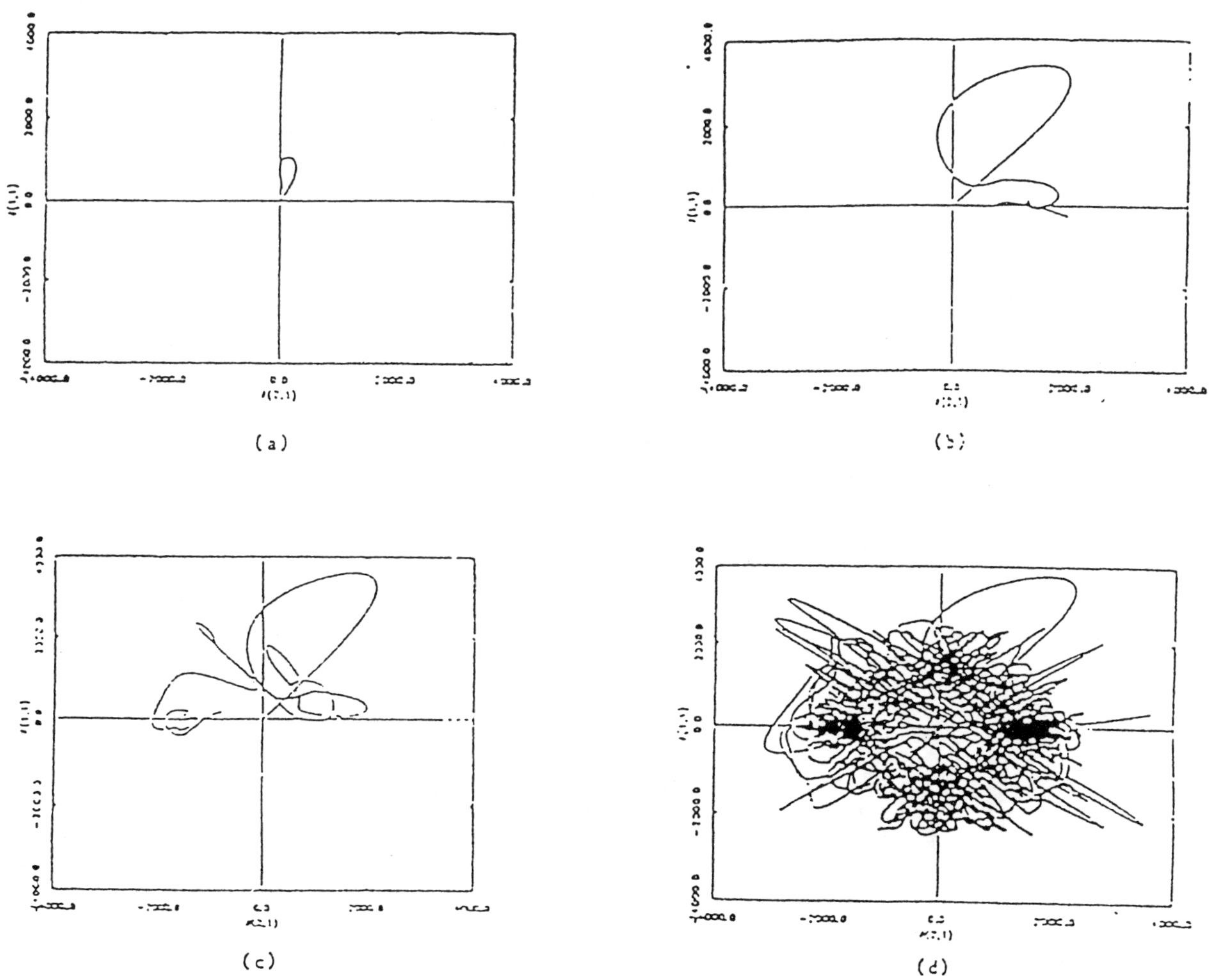

Figure 9.14 Numerical solutions of the 12-mode infinite Prandtl number equations projected onto the $\theta_{1,1}$, $\theta_{2,1}$ plane of the 12-dimensional phase space. (a) R = 10,000 (r = 15.21), (b) Ra = 30,000 (r = 45.62), (c) Ra = 43,000 (r = 65.39), (d) Ra = 45,000 (r = 65.39).

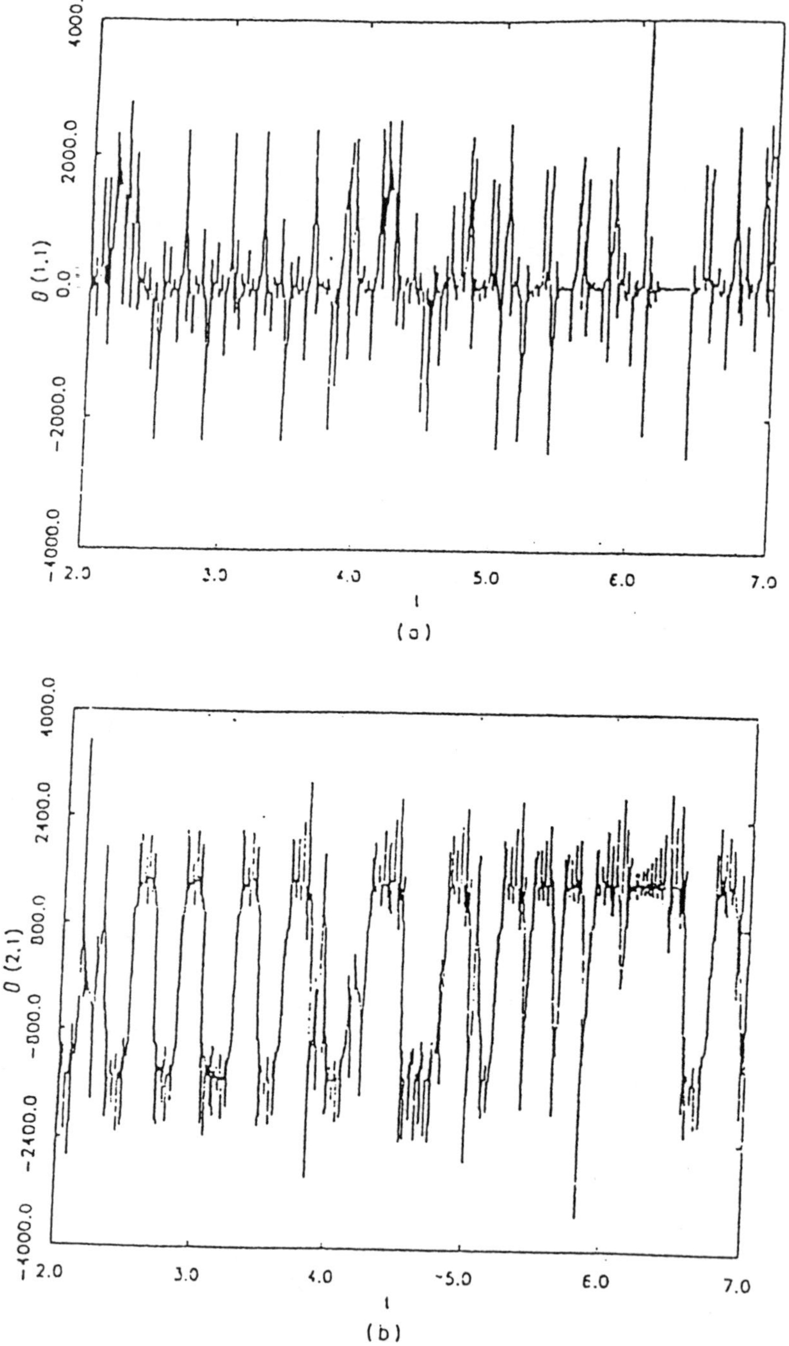

Figure 9.15 Time dependence of the coefficients (a) $\theta_{1,1}$ and (b) $\theta_{2,1}$ for the solution given in Figure 9.14d, Ra = 45,000.

(subcritical pitchfork plus two saddles) typically produces hysterisis effects when the saddle has one stable branch and one unstable branch. Here, the "$\theta_{1,1}$ mixed" solution has one unstable manifold (out of 12) on one side of the saddle, and two unstable manifolds on the other.

The second bifurcation of the conduction solution is at Ra = 1315.023, where two unstable symmetric fixed points dominant in the subharmonic $\theta_{2,1}$ mode appear. Since these fixed points contain no component in the $\theta_{1,1}$ mode, we call these "$\theta_{2,1}$ pure" unstable solution branches. Each of these branches becomes stable and undergoes a pitchfork bifurcation at Ra = 2041.918, resulting in the branches solution labeled "$\theta_{2,1}$ mixed" in Figures 9.16a and 9.16b. Then $\theta_{2,1}$ mixed mode branches nearly connect with the $\theta_{1,1}$ mixed mode branches.

Both the fundamental and the subharmonic pure mode solutions are stable between Ra = 2369 and Ra = 3.802×10^4. The trajectories in Figures 9.14a and 19.14b have the same initial condition, yet in Figure 9.14a it converges to the fundamental subharmonic stable fixed point. Presumably this is because the unstable mixed-mode branches disrupt the seperatix between the basin of attraction of the fundamental and subharmonic pure mode solutions. Note that the transition from Figure 9.14a to Figure 9.14b occurs at a Rayleigh number above the stability limit of the fundamental mode.

The third bifurcation of the conduction solution is at Ra = 4.140×10^4, where two unstable symmetric solutions in the fundamental mode $\theta_{1,1}$ appear. These are labelled "new $\theta_{1,1}$ pure" in Figure 9.16a. Each of these undergo Hopf bifurcations at Ra = 5.23×10^4 and Ra = 5.53×10^4. At no point does the origin itself undergo a Hopf bifurcation, nor does the conduction solution bifurcate to mixed mode solutions.

We detected no Hopf bifurcations for the conduction or fundamental harmonic solutions; however, the stable subharmonic branch undergoes two Hopf bifurcations (Figure 9.16b), one at Ra = 4.37×10^4 and one at Ra = 6.36×10^4. Each mixed-mode saddle ($\theta_{1,1}$ mixed) undergoes two Hopf bifurcations, at Ra = 4.491×10^4 and Ra = 5.039×10^4. Each mixed-mode saddle ($\theta_{2,1}$ mixed) undergoes six Hopf bifurcations. Each of these Hopf bifurcations sheds stable or unstable periodic orbits which are responsible for the oscillations of the trajectory at Ra = 4.5×10^4 shown in Figure 9.14d.

In Figure 9.17, the first 7000 points of the trajectory at Ra = 4.5×10^4 are projected onto the $\theta_{1,1}$, $\theta_{2,1}$ plane (dotted line) superimposed on the central portion of the branches of the fixed points (solid lines). Note that the trajectory weaves aperiodically around several Hopf bifurcations (circles).

Physically, infinite Prandtl number, high Rayleigh number convection becomes time-dependent through boundary layer instabilities that generate thermal plumes. In terms of spectral expansions, these instabilities result form the nonlinear coupling in the convective terms of the heat equation.

The 12-mode expansion gives solutions that are clearly chaotic. A valid question is whether higher order expansions also yield chaotic solutions. In order to examine this question 24-and 40- mode expansions have been considered. At Ra = 10^4, both models converge to fixed points, and the energy-distribution among the Fourier modes is qualitatively correct; modes which should be near-zero are near-zero; while large- amplitude modes have comparable amplitudes, within 10%. At Ra = 5×10^4, however, the temporal behavior of the 24-mode model and the 40-mode model diverge significantly; the 40-mode model goes to a fixed-point while the 24-mode model goes to a periodic orbit. At higher Rayleigh numbers, this relationship inverts, the 24-mode model going to a stable fixed-point, and the 40-mode model to a periodic orbit. Therefore, it is not always true that adding more modes in the expansion will always reduce the time-dependent behavior of the solutions - or the dimension of the attractor - in the infinite Prandtl case; sometime the opposite is true.

In Figure 9.18 we show the periodic orbit of the 40 mode system projected onto the $\theta_{1,1}$ - $\theta_{2,1}$ plane at Rayleigh number Ra = 1.9×10^5 . Because the system is symmetric under the transformations $\theta_{m,n} \rightarrow - \theta_{m,n}$ this periodic orbit also occurs in the other quadrants of the system. Figure 9,19 shows the periodic orbit after bifurcation to a 2-Torus. This occurs near Rayleigh number Ra = 1.95×10^5. A second temporal frequency has been introduced that causes the periodic orbit to wind around in two dimensions, remaining on the surface of a torus. The ratio between the two frequencies is approximately 50.

The study of thermal convection at infinite Prandtl number may be the best way to relate low-order chaotic systems to high-order turbulent systems. At infinite Prandtl number the viscous damping occurs in the

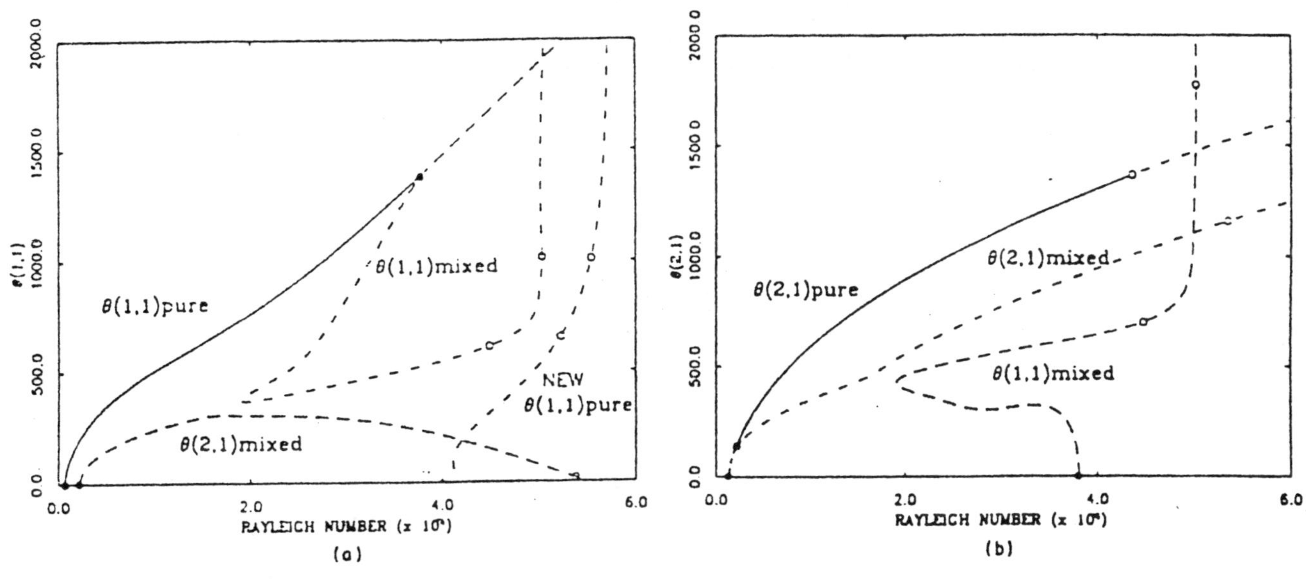

Figure 9.16. Bifurcation diagram for the 12-mode infinite Prandtl number equations. The fixed points of (9.3) are projected onto the (a) $\theta_{1,1}$, Ra plane (b) onto the $\theta_{2,1}$, Ra plane. Stable branches are shown as solid lines, instable branches as dahed lines, pitchfork bifucations as solid circles, and Hopf bifurcations as open circles.

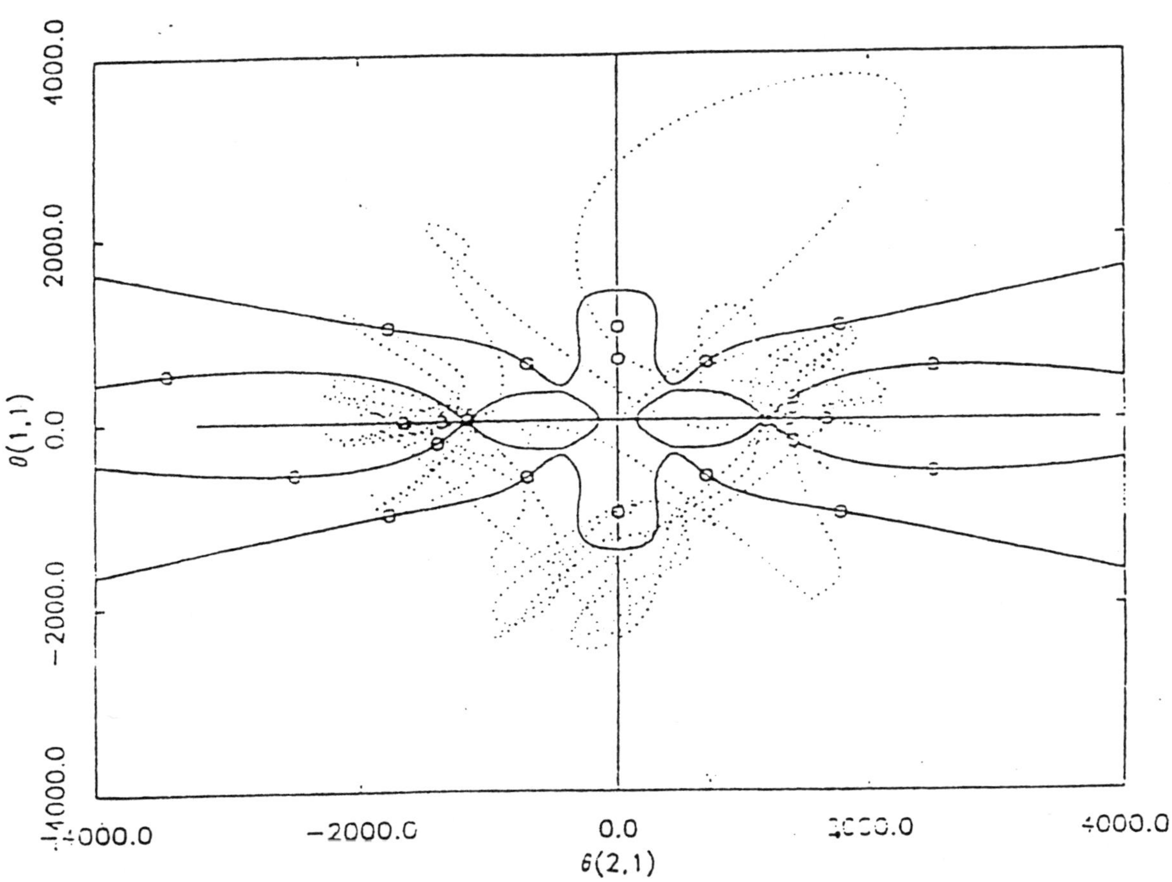

Figure 9.17 Loci of the fixed points from Figure 9.16 projected onto the $\theta_{1,1}$, $\theta_{2,1}$ plane. Superimposed as a dotted line is the time evolution of chaotic solution from Figure 9.14d.

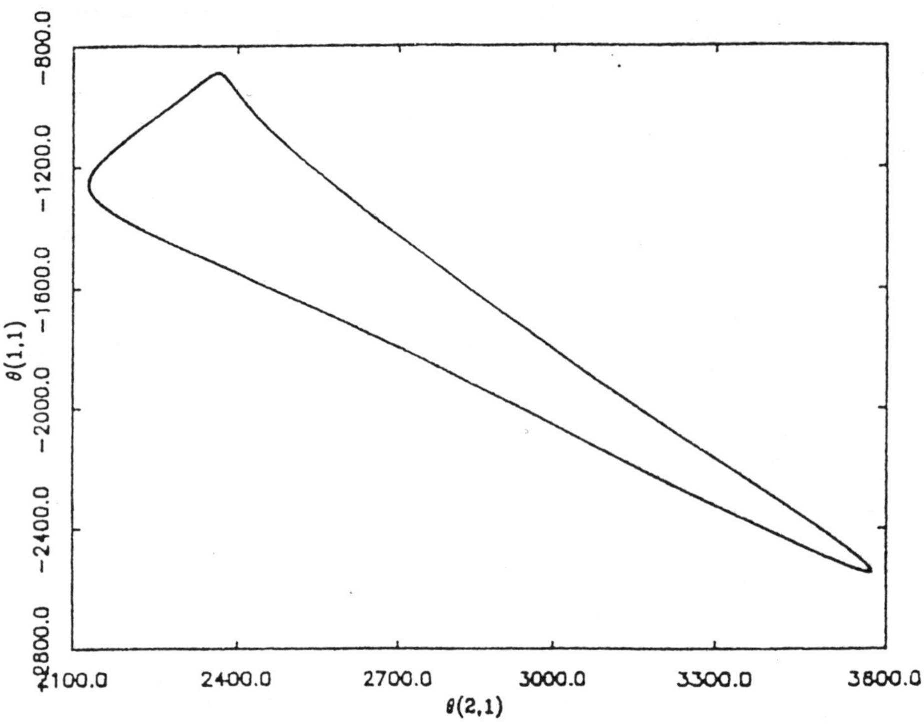

Figure 9.18 Periodic orbit of the 40-mode system at Ra = 1.9 x 10^5.

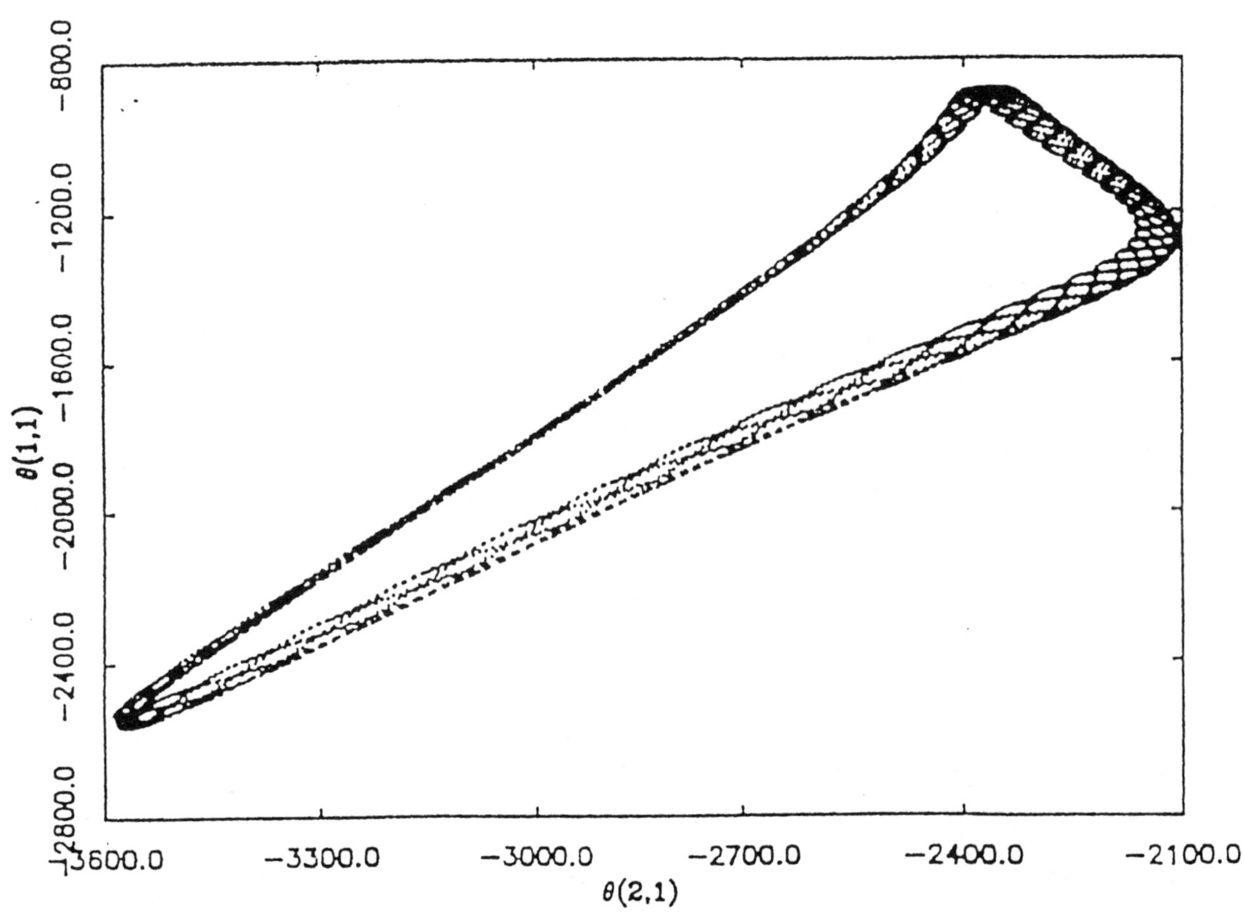

Figure 9.19. Periodic orbit after the transition to a 2-torous in the 40-mode system at Ra = 1.19 x 10^5.

same wavelength band as the buoyancy input. Flows have relatively little short wavelength energy. The 40-mode expansion considered here approaches the resolution of published numerical calculations that exhibit fluid turbulence (Christensen, 1987; Lennie et al., 1988; Hansen and Ebel, 1988; Vincent and Yuen, 1988; Christensen and Yuen, 1989).

9.4 Conclusions

We have examined two problems in solid earth dynamics. We have considered truncated expansions of the equations governing thermal convection at infinite Prandtl number. Our studies indicate chaotic behavior and are a first attempt to bridge the gap between low-order truncations and numerical calculations indicating turbulent solutions.

Our studies of slider-block models show classical chaotic behavior. It remains for future studies to develop the relations between these analogs models and distributed seismicity in the crust.

Acknowledgements: This research was supported by NASA grants NGT-50203 and NAG 5-860, and NSF grant EAR 8721172.

References

Bak, P. and C. Tang, 1989, Earthquakes as a self-organized critical phenomenon, J. Geophys. Res., v. 94, p. 15,635-15,637.

Bak, P., C. Tang and K. Wiesenfeld, 1988, Self-organized criticality, Phys. Rev., v. A38, p. 364-374.

Burridge, R. and L. Knopoff, 1967, Model and theoretical seismicity, Seis. Soc. Am. Bull., v. 57, p. 341-371.

Byerlee, J., 1978, Friction of rocks, Pure Ap. Geophys., v. 116, p. 615-626.

Cao, T. and K. Aki, 1984, Seismicity simulation with a mass-spring model and a displacement hardening-softening friction law, Pure Ap. Geophys., v. 122, p. 10-24.

Cao, T. and K. Aki, 1986, Seismicity simulation with a rate-and-state friction law, Pure Ap. Geophys., v. 124, p. 487-513.

Carlson, J.M. and J.S. Langer, 1989, Mechanical model of an earthquake fault, Phys. Rev., v. A40, p. 6470-6484.

Christensen, U.R., 1987, Time-dependent convection in elongated Rayleigh-Benard cells, Geophys. Res. Let., v. 14, p. 220-223.

Christensen, U.R. and D.A. Yuen, 1989, Time-dependent convection with non-Newtonian viscosity, J. Geophys. Res., v. 94, p. 814-820.

Cohen, S., 1977, Computer simulation of earthquakes, J. Geophys. Res., v. 82, p. 3781-3796.

Dietrich, J.H., 1972, Time-dependent friction as a possible mechanism for aftershocks, J. Geophys. Res., v. 77, p. 3771-3781.

Dietrich, J.H., 1981, Constructive properties of faults with simulated gouge, in Mechanical Behavior of Crustal Rocks, N.L. Carter, N. Friedman, J.M. Logan, and D.W. Steans, eds., pp. 103-120 , American Geophysical Union, Geophys. Mono. 24.

Gu, J.C., J.R. Rice, A.L. Ruina, and S.T. Tse, 1984, Slip motion and stability of a single degree of freedom elastic system with rate and state dependent friction, J. Phys. Solids, v. 32, p. 167-196.

Hansen, U. and A. Ebel, 1988, Time-dependent thermal convection for multiscale flow in the earth's mantle, Geophys. J., v. 94, p. 181-191.

Huang, J. and D.L. Turcotte, 1990, Are earthquakes an example of deterministic chaos?, Geophys. Res. Let., v. 17, p. 223-226.

Ito, K. and M. Matsuzaki, 1990, Earthquakes as self-organized critical phenomena, J . Geophys. Res., v. 95, p. 6853-6868.

Keilis-Borok, V.I., 1990, The lithosphere of the earth as a nonlinear system with implications for earthquake prediction, Rev. Geophys., v. 28, p. 19-34.

Keilis-Borok, V.I. and V.G. Kossobokov, 1990, Premonitory activation of earthquake flow: Algorithm M8, Phys. Earth Planet Int., v. 61, p. 73-83.

Keilis-Borok, V.I. and I.M. Rotwain, 1990, Diagnosis of time of increased probability of strong earthquakes in different regions of the world: Algorithm CN, Phys. Earth Planet. Int., v. 61, p. 57-72.

Lennie, T.B., D.P. McKenzie, D.R. Moore and N.O. Weiss, 1988, The breakdown of steady convection, J. Fluid Mech., v. 188 , p. 47-85e.

Lorenz, E.N., 1963, Deterministic nonperiodic flow, J. Atmos. Sci., v. 20, p. 130-141.

Molchan, G.M., O.E. Dmitrieva, I.M. Rotwain and J. Dewey, 1990, Statistical analysis of the results of earthquake prediction, based on bursts of aftershocks. Phys. Earth Planet. Int., v. 61, p. 128-139.

Nakanishi, H., 1990, Cellular-automaton model of earthquakes with deterministic dynamics, Phys. Rev., v.A41, p. 7086-7089.

Nussbaum, J. and Ruina, A., 1987, A two degree-of-freedom earthquake model with static/dynamic friction, Pageoph, v. 124, p. 629-656.

Rice, J.R. and S.T. Tse, 1986, Dynamic motion of a single degree of freedom system following a rate and state dependent friction law, J. Geophys. Res., v. 91, p. 521-530.

Ruina, A., 1983, Slip instability and state variability laws, J. Geophys. Res., v. 88, p. 10,359-10,370.

Rundle, J.B. and D.D. Jackson, 1977, Numerical simulation of earthquake sequences, Seis. Soc. Am. Bull., v. 67, p. 1363-1377.

Saltzman, B., 1962, Finite amplitude free convection as an initial value problem, J. Atmos. Sci., v. 19, p. 329-341.

Schreider, S. Yu, 1990, Formal definition of premonitory seismic quiesence, Phys. Earth Planet. Int. v. 61, p. 113-127.

Sornette, A. and D. Sornette, 1989, Self-organized criticality and earthquakes, Europhys. Let., v. 9, p. 197-202.

Sornette, A. and D. Sornette, 1990, Earthquake rupture as a critical point:

Consequences for telluric precursors, Tectonophys., v. 179, p. 327-334

Stewart, C.A. and D.L. Turcotte, 1989, The route to chaos in thermal convection at infinite Prandtl number 1. Some trajectories and bifurcations, J. Geophys. Res., v. 94, p. 13,707-13,717.

Takayasu, H. and M. Matsuzaki, 1988, Dynamical phase transition in threshold elements, Phys. Let., v. A131, p. 244-247.

Tullis, T.E. and J.D. Weeks, 1986, Constitutive behavior and stability of frictional sliding of granite, Pure Ap. Geophys., v. 124, p. 383-414.

Vincent, A.P. and D.A. Yuen, 1988, Thermal attractor in chaotic convection with high-Prandtl-number fluids, Phys. Rev., v. A38, p. 328-334.

Vincent, A.P. and D.A. Yuen, 1989, Multimode mean-field filtering experiments for chaotic thermal convection, Geophys. Astrophys. Fluid Dyn., v. 47 , p. 131-156.

A1

APPENDIX 1

CHAOSSARY (A SHORT GLOSSARY OF CHAOS)

Gerard Middleton
Department of Geology
McMaster University
Hamilton, Ontario
L8S 4M1

Aggregation

When two particles collide and stick together (without completely merging) they are said to aggregate. A particular form of aggregation that has been much studied is diffusion-limited aggregation (DLA: first named by Witten and Sander in 1981, see Sander, 1987). In DLA a seed particle grows by aggregation of other particles approaching it by a random-walk diffusive process. The aggregates that result have fractal properties and resemble certain natural dendrites: their study has also thrown light on the formation of skeletal crystals (such as snow flakes). Similar structures are formed by quite different physical processes, such as viscous fingering in porous media, and chemical dissolution. See also Multifractal.

Arnol'd Tongues

see circle map.

Attractor

The term is used quite generally to describe the state towards which dissipative dynamic systems (and their mathematical analogues) tend to converge. A pendulum with friction, for example, tends to converge on the position of rest, with the bob hanging vertically below the support. A pendulum with friction which is periodically given an impulse may tend to converge on a particular cycle of motion, called a limit cycle (a simple equation to describe such a pendulum or oscillator is called the van de Pol equation, and has application in electronics). These types of attractors have been known for many years, but a new type of attractor, now called a Strange Attractor, was described for the first time by a meteorologist, Lorenz, in 1963.(Lorenz, however, did not use the term itself. It is not clear who did originate the term: several authors attribute it to Smale, 1967, but though he recognized many of the properties of strange attractors, and certainly refered to "attractors" in his 1967 paper, the term "strange attractors" does not seem to appear in the paper. Ruelle, 1980, writes that he asked Takens if he had originated it, in Ruelle and Takens, 1971, and he replied "Did you ever ask God if he created this damned universe...I don't remember anything...")

Attractors are generally portrayed either in phase (or state) space or in planes cut through phase space. These planes are called Poincaré sections.

Strange Attractors are characteristic of certain dissipative dynamic systems of more that two dimensions, and describe a particular kind of chaotic behaviour. Their most characteristic feature is that, although the attractor is confined within a certain region of phase space, trajectories within the attractor show "sensitivity to initial conditions". This means that trajectories that begin from positions very close together diverge very rapidly (in fact, exponentially) so that their positions in phase space are soon very far apart. A measure of the rate of divergence is provided by the Liapunov exponent. Divergence takes place even though both trajectories are constrained to lie within, or very close to, the attractor. Divergence of trajectories may be thought of as a stretching process, because a group of points originally confined within a small cube, for example, commonly become "stretched out" and are confined within a thin plate-like volume down trajectory. Divergence of trajectories may be reconciled with confinement, by the process of "folding" trajectories back on themselves in phase space.

Some (but perhaps not all) strange attractors show fractal properties.

Basin of Attraction

In dissipative dynamic systems, trajectories in phase (or state) space tend to converge on one or more attractors. In systems with more than one attractor, it is possible to map out (theoretically, or more usually by computation) which initial conditions (regions in phase space) lead to which attractors. These regions constitute the basin of attraction of a particular attractor. The boundaries between basins have extremely complex geometry in many dynamic systems, and constitute very striking fractals. A well-known example is the boundary between trajectories that diverge to infinity and those that converge in the map

A3

$$z_{n+1}^2 = z_n^2 - c$$

where z and c are complex numbers. If c is a constant, and the initial value, z_0, is plotted on the screen (using different colours or shades depending on the rate of convergence or divergence) then the set of points that do not converge is called a Julia set. If the initial point is fixed at $z_0 = 0$, and c is plotted on the screen (using different colors for different rates of divergence) then the set of points that do not converge is called the Mandelbrot set.

Bifurcation

The general phenomenon being described is the transition from a single-valued function to a double-valued function. For example, the well-known logistic map:

$$x_{n+1} = k\, x_n(1 - x_n)$$

is a single valued function for values of $k < 3$, but becomes double valued in the range $3 < k < 3.5$. At higher values of k, further bifurcations are observed, and the function becomes 4-valued, 8-valued, etc. This is an example of period doubling and of one particular type of bifurcation of interest to students of chaos, called a pitchfork (or period-doubling) bifurcation. According to Arnol'd (1986, p.2) bifurcation "...is used in a broad sense for designating all sorts of qualitative reorganizations or metamorphoses of various entities resulting from a change of the parameters on which they depend."

A general classification of bifurcations is possible (e.g., Bergé et al., 1984, Appendix A; Thompson and Stewart, 1986, Chapter 7). The topic is closely related to the classification of "catastrophes" described by catastrophe theory. The following simplified discussion is taken from Glass and Mackey (1988).

The number and/or stability of steady states or cycles which a system displays may depend on the value of one or more parameters (in the example above, the parameter k). Any such change is described as a bifurcation and the critical value of the parameter(s) define a bifurcation point. To understand the stability of a particular state (point in the system), it is useful to consider a small region around that point. In this region it is generally possible to simplify the non-linear equations governing the system to a set of linear equations. The stability of sets of linear equations is well understood and may be found from the properties of the Jacobian of the system. For steady states in two dimensions, the possible conditions of stability may be described (after Poincaré) as one of the following: (i) focus -- paths spiral in towards the steady state; (ii) node -- paths move more directly in towards the steady state; (iii) saddle point -- some paths move towards and some away from the steady state. These different conditions can be recognized by examining the eigenvalues of the Jacobian. Forms of instability, or bifurcations, can also be recognized in this way. For example, if the system is characterised by two complex eigenvalues, then the Hopf Bifurcation takes place as the eigenvalues cross the imaginary axis, i.e., as the real part of the complex eigenvalues changes from negative to positive values. Such bifurcations may be further classified as supercritical, if a single steady state is relaced by a limit cycle whose amplitude gradually grows larger, or subcritical if the single steady state is suddenly relaced by a large amplitude limit cycle.

Brusselator

A term applied informally to a system of differential equations that results from a particular set of chemical reactions:

$$\begin{aligned} A &\Rightarrow X \\ 2X + Y &\Rightarrow 3X \\ B + X &\Rightarrow Y + D \\ X &\Rightarrow E \end{aligned}$$

It is discussed by Prigogine (1980) and is named for the "Brussels school" of thermodynamics, which he leads. The system gives rise to a limit cycle.

Cantor Dust (or Set)

A fractal produced by subdividing a line into parts, deleting some of them, and repeating the process indefinitely. Usually the line is divided into three equal parts, and the central part is deleted: in this case the fractal dimension is 0.63...

Catastrophe

The term "catastrophe theory" was introduced by E.C. Zeeman, to describe a theory originally proposed by the French mathematician, Rene Thom. "Catastrophes are abrupt changes arising as a sudden response of a system to a smooth change in external conditions." (Arnol'd, 1986, p.2). An elementary discussion is given by Zeeman (1976). Catastrophe theorists recognize seven "elementary" (types of) catastrophes. The theory is disavowed by many chaos theorists (e.g., Arnol'd, Smale) because of the highly speculative, qualitative applications made by Zeeman and his school to the social sciences.

A5

Chaos

Some authorities doubt that a satisfactory definition of chaos can be given in the present (?chaotic) state of knowledge. The word comes from the Greek meaning empty space or void, and is commonly used to imply the absence of any order or structure. Its first use in the modern sense is perhaps that by Li and Yorke (1975).

The present use of the term in mathematics and physics is closely connected with the behaviour observed in certain relatively simple, non-linear, dynamic systems, or sets of ordinary differential equations. Under some conditions, such equations not only display instability, but unstable behaviour of a particularly complex kind. Probably the single most important feature of these systems is that "precise knowledge of the past evolution of a system over an arbitrarily long time does not aid in predicting its subsequent evolution past a limited time range" (Bergé et al., 1986, p.103). Some chaotic, dissipative systems of three or more dimensions also display the phenomenon of strange attractors.

Circle Map

A map of the form

$$\Theta_{n+1} = \Theta_n + \Omega - Kg(\Theta_n)$$

where Θ is a circular measure of period one (so varies from 0 to 1 rather than the usual 2π radians), and Ω (omega) is the frequency in the absence of the nonlinear term $g(\Theta)$. The nonlinear term, for example, might be a sine function (giving the Sine Circle Map). K is a coefficient which determines the strength of the nonlinear component. Thus circle maps are a whole class of nonlinear difference equations that map a point on a circle to another point on the circle. The average number of rotations per iteration is called the winding number: if there is no nonlinear term (K=0) then the winding number is simply equal to Ω.

Circle maps can be used to represent the dynamics of a system with two frequencies, for example, a damped pendulum with a natural frequency (normalized to one) forced by a mechanism whose driving frequency (1+f) differs from the natural frequency. At small values of f, the motion is quasi-periodic. At larger values phase locking occurs, and the motion is periodic. As the strength of the nonlinear component increases, the system may become chaotic.

The behaviour of the oscillating system can be portrayed by plotting K against Ω. For a given K, there will be some values of Ω for which there is phase locking between the two frequencies: for these values there is resonance even if the initial ratio between the two frequencies is not exactly rational. The range of Ω values for which this is possible increases as K increases. Lines can be drawn on the diagram separating these regions from regions in which phase locking does not take place: the phase-locked regions are called Arnol'd tongues.

Devil's Staircase

A name given to a continuous "curve" composed of an infinite number of steps. A simple example can be constructed from a Cantor dust construction. Cantor dust is produced by subdividing a line, removing part, and repeating the process indefinitely. If a bar is substituted for the line, then the process can be modified, for example, by requiring that the mass of the original bar is conserved as parts are removed: this is achieved by proportionately increasing the mass of the remaining parts. Suppose the original bar extended from $x = 0$ to $x = 1$, then a graph showing the way the cumulative mass of the pieces, $M(x)$ as a function of x, would be a Devil's Staircase (Feder, 1988, p.69). The process of concentrating an originally uniform mass distribution onto many small regions with a fractal distribution has been called "curdling" by Mandelbrot.

The Devil's Staircase may be used to represent the dynamics of driven oscillators, and similar dynamic systems (Bak, 1986). In this case the level surfaces of the "steps" represent rational ratios between the frequencies of the oscillator and its driver produced by phase locking (resonance between the natural frequency of the oscillator and the initial frequency of its driver).

Difference Equation

An equation that relates the value of a function $f(x,t_{i+1})$ to the value of the function at an ealier time t_i , where the difference between the two times is finite. Difference equations may be studied in their own right, and arise naturally in certain branches of science (for example, in studies of sucessive populations which arise from a regular birth/death cycle) but generally they result from recasting differential equations in finite difference form, so that they may be solved numerically.

Difference equations show more chaotic behaviour than their corresponding differential equation. An example of this is the logistic equation, which in its differential form has an exact integral (and is therefore non-chaotic) whereas in its difference form it is the classic example of a simple one-dimensional map showing chaos.

Dimension

The common concept of dimension is the topological dimension, that is, the dimension of the space needed to represent an object. For a point this is zero, for a line one, for a plane 2 and for a sphere or cube 3. It is always an integer. A second concept is that of fractal dimension. Various definitions of this are available: see fractal and Koch curve. A common one is the Hausdorf-Besicovitch dimension, which can be defined as follows: suppose we have a set of points in an n-dimensional space. We try to cover the points by a set of N hypercubes, each of linear dimension ε, where N is the smallest possible number of cubes necessary to cover all the points. Then the H-B dimension D is defined by:

$$D = \lim_{\varepsilon \to 0} \left(\frac{\ln N(\varepsilon)}{\ln(1/\varepsilon)} \right)$$

> "The Hausdorff dimension of a set is obtained from covering that set minimally with open balls (intervals in one dimension) with radii that may be different. The limit capacity involves the same process except that one is restricted to balls of equal radii. In both cases the maximum radius is allowed to go to zero yielding a measure for the set by summing over the ball diameters raised to the power D. The greatest lower bound of the powers {D} causing that measure to be zero defines the Hausdorff dimension or the limit capacity, as the case may be." (Essex and Nerenberg, 1990).

Duffing's Equation

A differential equation for a nonlinear driven oscillator, originally developed for applications in electronics. The nonlinear term is cubic (see oscillator). It shows a wide range of chaotic behaviour (e.g., see Holden, 1986).

Eigenvalue and Eigenvector

Eigenvalues are characteristic values of some mathematical expression. The term is generally applied to the characteristic roots of a square matrix. If the matrix is **A**, the eigenvalues λ are the roots of the polynomial:

$$|\mathbf{A} - \mathbf{I}\lambda| = 0$$

where I is the unit matrix, and the vertical lines indicate the determinant of the matrix between them. Eigenvalues have many application, from factor analysis in statistics to determining the relative magnitudes of the principal stresses or strains in rock mechanics or structural geology. They are used in determining the stability of linear differential equations (see Jacobian), and in the classification of different types of bifurcations.

The eigenvectors are vectors in space associated with each eigenvalue. In the case of a stress matrix, the eigenvectors indicate the directions of the principal stresses, and the eigenvalues their relative magnitude.

Ergodic Hypothesis

This is a phrase commonly applied to time series: it supposes an equality between "ensemble" averages (the average of all possible time series) and averages, over time, of a single time series. The phrase is also used in a more general sense in thermodynamics. A simple statement given by Maxwell (see Prigogine, 1980) is: "The only assumption which is necessary for a direct proof of the problem of thermodynamic equilibrium is that the system, if left to itself, will sooner or later pass through every phase which is consistent with the equation of energy." The ergodic hypothesis about random systems with many degrees of freedom may be contrasted with the usual hypothesis in classical mechanics, that simple systems whose initial conditions and equations of motion are fully known, will exhibit only completely predictable behaviour, with trajectories confined to a small part of phase space.

It has now been shown (see KAM) that many systems are intermediate in character: they show well defined regions of predictable deterministic behaviour, and other regions within which there is chaotic behaviour.

Expectation

This has a technical meaning in statistics. The expected value of a function of a continuous variable, g(x) is defined as

$$E(g(x)) = \quad g(x)p(x)\,dx$$

where p(x) is the probability density function of x. If g(x) = x then the expectation of x, otherwise called the expected value of x, correspondance to the average value of x. If x is a discrete rather than continuous variable, then we replace the integral sign with a sum, and we can think of the probabilities as being the "weights" associated with each value of x (or with each value of the function g(x)).

Feigenbaum Constant

In the sequence of bifurcations observed in the logistic equation
Feigenbaum noted in 1975 that bifurcations took place at values of k of 3.000, 3.449..., 3.544..., etc., a sequence of values which appear to converge on a limit 3.569946... He further noted that the value of k at the nth bifurcation is given by

$$k_n = 3.569946... - 2.6327(4.669202...)^{-n}$$

A similar phenomenon is noted in examples of period doubling bifurcations observed in many other many other iterative maps (see Poincaré section). Though the other numbers vary from map to map the value F = 4.669202... does not: it is therefore a "fundamental constant" similar to pi, e, etc. and it is now called the Feigenbaum constant. A simplified "explanation" of this constant is given in Holden (1986, p.50-51). Feigenbaum has shown that the universal properties of bifurcations are a consequence of renormalization group theory. See Chapter 5.

Fixed Point

Also called an "equilibrium point", or a "steady state". The term "singular point" is also used by some authors (Edelstein-Keshet, 1988, p.176). A point in phase space for which

$$x = f(x)$$

in other words, the solution of the (set of) differential equation(s) at this point (value of x) is the value of x itself (where x may be a vector). The concept is easier to understand for a difference equation

$$x_{n+1} = f(x_n)$$

where it implies that x(n+1) = x(n), i.e., iteration does not change the value of x. If the iteration corresponds to taking the next intersection of a trajectory with a Poincaré section of a phase space, then the fixed point corresponds to a limit cycle in the phase space.

Fourier Series

An infinite series consisting of a sum of sine and cosine terms: therefore, the series has a period 2π. The numerical coefficients of the sine and cosine terms can be chosen so that virtually any wave form can be represented to a high degree of approximation. In Fourier analysis, also called spectral analysis a time series is analysed to determine the relative contribution to the total variance of the variable x(t) contributed by each of the frequencies needed to approximate it by a Fourier series. If the time series has strong, regular period components, then these are revealed as strong peaks in the power spectrum. If the data is completely random, then the spectrum shows no peaks. Spectra of many 'chaotic' or 'turbulent' time series show no well-defined peaks, but a systematic decrease in the power (contribution to the total variance) of progressively higher frequency components. A time series x(t) may be converted by Fourier analysis into a function of frequency y(f) rather than of time. x(t) and y(f) are then said to be a pair of Fourier transforms.

A10

An example of the application of Fourier series to partial differential equations is provided by a vibrating string. The motion of the string is described by a partial differential equation, and the solution may be represented by a Fourier series. The first terms represent the fundamental vibration frequency of the string, and higher terms represent the higher harmonics. For a complete representation, an infinite number of terms would be required, therefore, it may be said that the partial differential equations has an infinite number of degrees of freedom. But in practice, a very good representation may be obtained using only the first few terms of a Fourier series. Very high order terms may be rapidly damped out by friction. Similarly, phenomena such as thermal convection in fluids may be satisfactorily represented in some cases by only a few terms of their Fourier expansion. See Chapters 3 and 9.

Fractal

This term was introduced by Mandelbrot in 1975 and has been defined (even by him) in different ways. One version given by Feder (1988, and attributed by him to Mandelbrot) is

> "A fractal is a shape made of parts similar to the whole in some way"

The topic is a large one, and is related to chaos by the fact that many chaotic systems display iterative mappings (see <u>Poincaré section</u>) which have fractal properties. This means that small parts of the mapping, when enlarged, closely resemble the parts of the original mapping (at the larger scale) and that this process can be continued over several scales of enlargement.

An important property characterizing fractals is their <u>dimension</u>. Various different definitions of fractal dimension have been proposed, with the most common ones being the similarity dimension, the box-counting dimension, and the Hausdorff (or <u>Hausdorff-Besicovitch</u>) dimension. Rigorous definitions can be found in Feder (1988) or Mandelbrot (1982). The following heuristic introduction is taken from Sorensen (1984).

A line has one dimension and a flat plane two dimensions, and so on (this is the <u>topological dimension</u>, the dimension of the space necessary to display a line or a plane). An irregularly shaped line may be considered to have a dimension somewhere between 1 and 2: the dimension increases as the line becomes more irregular, and only reaches 2 when the line is so irregular that it fills the entire plane. To measure the fractal dimension of a line (the classic example is an irregular, rocky coastline) we measure the length of the coastline at several different scales. We do this by "walking" a pair of dividers (or compass, i.e., a unit length ε) along a map of the coastline. If we then plot the logarithm of the total length against the logarithm of ε, we find that we define a straight line (this was actually first done by L.F. Richardson in 1961). The line corresponds to the equation:

$$L(\varepsilon) = a\,\varepsilon^{-D}$$

where D is a fractal dimension. D has a value between 1 and 2. The concept may be generalized to higher dimensions. For example, an irregular surface may have a fractal dimension between 2 and 3, and so on. Note that not every irregular line or surface has a (constant) dimension, but only those with fractal properties. Though most fractals have non-integer dimensions, a fractal dimension may also be an integer. An example of a fractal with an integer dimension is the Peano curve, which (in the limit) completely fills a plane and so has dimension 2.

Since most (?perhaps all) strange attractors have fractal properties, numerical techniques have been developed to measure the fractal dimension of these attractors (e.g., Holden, 1986, Chapter 14). It has also been found that many strange attractors and other fractals (such as eddies in turbulent fluids) are not adequately described by a single dimension, but require a spectrum of dimensions (see multifractal).

True fractals are mathematical objects (sets) which can be fully explored only by using the concepts of pure mathematics, in fields such as set theory and topology. The properties of fractals can be considered to define a type of geometry, different from ordinary (Euclidean or Cartesian) geometry. Real objects resemble fractals only within a certain range (see scale); of course, it is equally true that real objects can be described by ordinary geometry only within a certain range (e.g., scales at which matter can be considered continuous and "smooth") and to a certain degree of approximation.

Hamiltonian System

For a mechanical system, the Hamiltonian is a function of the coordinates and momenta of the parts of the system, and is equal to the total potential and kinetic energy of the system.

Systems are described as Hamiltonian systems if the Hamiltonian is a constant with time. In mechanics, frictionless (non-dissipative) systems are Hamiltonian. These are also systems in which the force fields may be described by potentials, that is they may be expressed as the negative gradient of a scalar function (called the potential). In such a force field, the work done in moving a mass from one point to another does not depend on the path along which movement takes place. See also KAM, and Liouville's Theorem.

A12

Heaviside Function

A Heaviside Function, H(t) is equal to 0 if t is less than 0, and is equal to 1 if t is greater than or equal to one.

Hénon Map (or Attractor)

A dynamical system described by the following two-dimensional mapping:

$$x_{n+1} = 1 + y_n - Ax_n^2$$
$$y_{n+1} = Bx_n$$

First described by Hénon (1976).

Hénon Quadratic Map

A Hamiltonian (area preserving) two-dimensional map described by:

$$x_{t+1} = x_t \cos\alpha - y_t \sin\alpha + x_t^2 \sin\alpha$$

$$y_{t+1} = x_t \sin\alpha - y_t \cos\alpha + x_t^2 \cos\alpha$$

First described by Hénon (1969)

Hénon and Heiles Map

One of the first area-preserving maps known to give rise to chaos. It is generated as a Poincaré section of the following Hamiltonian equations:

$$H = \frac{1}{2}\sum_{i=1}^{2}(p_i^2+q_i^2)+q_1^2q_2-\frac{1}{3}q_2^3$$

$$\frac{\partial p}{\partial t} = -\frac{\partial H}{\partial q}, \qquad \frac{\partial q}{\partial t} = \frac{\partial H}{\partial p}$$

A13

First described by Hénon and Heiles (1964)

Intermittency

A signal is called intermittent "...if it is subject to infrequent variations of large amplitude." (Bergé et al., 1986, p.223). Intermittency is typical of transitions to turbulence in fluids, for example, velocity fluctuations measured close to the boundaries of turbulent boundary layers, jets or wakes. See Chapter 5.

Iterative (or Iterated) Map

A form of difference equation, where the next value $f(x_{n+1})$ of a function that depends on both x and t is calculated from the previous value $f(x_n)$. A classic example is the difference form of the logistic equation (see bifurcation).

Jacobian

If we have a system of two differential equations such that

$$dx/dt = f(x,y)$$

$$dy/dt = g(x,y)$$

then the Jacobian is the matrix

$$J = \begin{pmatrix} \frac{\partial f}{\partial x} & \frac{\partial f}{\partial y} \\ \frac{\partial g}{\partial x} & \frac{\partial g}{\partial y} \end{pmatrix}$$

If (x,y) are the coordinates of a fixed point then the properties of the eigenvalues of J can be used to determine the stability of the fixed point. The concept is easily generalized to more than two dimensions. The basic idea is that the partial derivatives define the slopes towards or away from the fixed point in the immediate vicinity of that point in phase space, and that these slopes define the stability of the point.

A14

KAM Theorem

This is a theorem in mathematics first postulated by Kolmogorov in 1954, and proved by Arnol'd (1963) and Moser (1962). It seems to have been first called the KAM theorem by Walker and Ford (1969: see that paper, or Lichtenberg and Lieberman, 1983, for original references). The basic idea is that Hamiltonian dynamical systems are not necessarily ergodic, that is, a trajectory does not necessarily "explore the entire region of phase space that energetically available to it" (Lichtenberg and Lieberman, 1983, p.3). In an ergodic system, the trajectories are entirely chaotic. The theorem thus "set the stage" for further investigations of chaos in dynamic systems, because many real systems are neither entirely regular, nor entirely chaotic (stochastic). From a combination of theory and computation

> "...an extraordinary picture of the phase space for weakly perturbed systems has emerged. The invariant surfaces break their topology near resonances to form island chains. Within these islands the topology is again broken to form yet other chains of still finer islands. On an even finer scale, one sees islands within islands. But this structure is only part of the picture, for densely interwoven within these invariant structures are thin phase space layers in which the motion is stochastic...For weakly perturbed systems with two degrees of freedom, the KAM surfaces isolate the thin layers of stochasticity from each other...as the perturbation strength increases a transition can occur in which the isolating KAM surfaces disappear and the stochastic layers merge, resulting in globally stochastic motion that envelopes the phase space. Often the phase space divides into three regions. One region contains primarily stochastic trajectories. It is connected to a second, mainly stochastic region containing large embedded islands. A third region consists of primarily regular trajectories and is isolated from the first two by KAM surfaces." (Lichtenberg and Lieberman, 1983, p.3-4).

Good examples of the phenomena described above are shown by the Hénon quadratic map and the Hénon and Heiles map.

Note that the KAM theorem applies only to systems which are conservative, therefore have no attractors.

Koch SnowFlake (or Curve)

This is one of the classic fractal curves, proposed by von Koch around 1904 (fide Voss, 1988). To construct a Koch curve an original straight line segment (for example, a side of an equilateral triangle) is replaced by a set of straight line segments, for example, the line segment is divided into three parts, and the central part is replaced by two lines equal in length to the original segment. The initial shape (the triangle in the example given) is called the initiator and the set of line segments that replace any original line segment, the generator. The process is carried out for all straight line segments in the initiator, and then repeated again for all the straight line segments in the transformed initiator, and so on indefinitely. If the process is carried through only a finite number of times, then the figure that results is not a true fractal, but a prefractal. In the equilateral triangle example, each straight line segment is replaced by four new segments (N = 4), each having one third of the original length (r = 1/3). The fractal (or similarity) dimension, D is then given by:

$$N = 1 / r^D$$

$$D = \log N / \log (1/r)$$

where N in the number of segments, and r is their length, relative to the original (i.e., N is the number of segments at the (n+1)th step of the generating process, divided by the number at the nth step). In the example

$$D = \log 4 / \log 3 = 1.26...$$

Note that Koch curves are a whole class of curves (with different fractal dimensions) that may be generated by such algorithms. They are continuous, but not differentiable at any point. They are also self-similar in the sense that, at any stage of the generation process, each small segment reproduces exactly a section of the larger curve at some previous stage.

Legendre Transformation

This is a transformation between two forms of a partial differential equation. For example, in dynamics the momenta, p, may be expressed a function of the coordinates, q, the velocities, u, and time, t. To express u in terms of the other variables, we make use of a Legendre Transformation (Percival and Richards, 1982). The transformations are also commonly used in thermodynamics. They are used in obtaining the $f(\alpha)$ function that characterizes multifractals.

Liapunov (or Lyapunov) Functions and Coefficients

Suppose that the behaviour of a system may be described by trajectories in phase (or state) space. Certain points or surfaces in the space may be stable eqilibrium points or surfaces, that is, trajectories beginning near these points rapidly converge on them. Others may be characterised by unstable equilibrium, and so on. Certain functions, $V(\mathbf{x})$, where

x is the vector of variables defining the phase space, may have the property that they describe smooth curves or surfaces along which the derivative of V with repect to time is always less than or equal to zero (remember that though all variables used to construct a phase space are functions of time, time is not one of the variables represented directly by the space). These functions are called Liapunov functions. Liapunov functions are useful in describing stability conditions in phase space.

For simplicity consider that the initial position in a Poincaré section through a phase space can be designated by a single variable, x. Then if the trajectory giving rise to this point by intersection with the Poincaré section is unstable, and we choose another trajectory separated from the first by a small distance dx, we expect that, as the trajectories are traced through time (corresponding to successive intersections on the Poincaré section) the distance dx will increase. If the increase is exponential we can represent dx for the nth iteration as

$$dx(n) = e^{n\lambda}dx(0)$$

where λ is defined as the Liapunov exponent. The size and sign of the exponent is therefore a measure of the instability of the trajectory. Strange attractors, which show sensitivity to initial conditions, have positive Liapunov exponents. A simple discussion is given in Holden (1986, Chapter 13).

Likelihood

This has a technical meaning in certain branches of statistics, (as in the method of maximum likelihood). The term is also used as a rough equivalent of probability.

Limit Cycle

In phase (or state) space, a limit cycle is any simple, closed trajectory that does not contain any singular points (i.e., there are no sigular points actually on the trajectory; there may be singular points inside the limit cycle trajectory). "Simple" implies that the trajectory cannot cross itself. Trajectories close to the cycle either outside or inside the limit cycle may converge on the limit cycle itself (which implies that the system is dissipative); if they converge on the cycle with increasing time then it is stable, if they diverge from it the cycle is unstable (Edelstein-Keshet, 1988, p.311-312). Note that a closed trajectory within a phase space is not a limit cycle unless there is local convergence towards or divergence from this trajectory.
Stable limit cycles are a form of attractor.

A17

Liouville's Theorem

This theorem states that for Hamiltonian systems, a small volume remains the same along trajectories in phase space. This expresses the "incompressibility of the flow" in phase space.

Logistic Equation

An equation describing growth of a some property x that is proportional to the product of x and the difference of x from some limit (e.g., 1)

$$dx/dt = rx(1-x)$$

It is a simple example of a nonlinear differential equation. The difference form of the logistic equation, called the logistic map

$$x_{n+1} = r\, x_n(1-x_n)$$

displays bifurcations for $r > 3.0$ and chaotic behaviour for some ranges of $r > 3.57$. See also Feigenbaum number.

Lorenz attractor (or equations)

see attractor. The equations are

$$\begin{aligned} dx/dt &= \sigma(y-x) \\ dy/dt &= rx - y - xz \\ dz/dt &= xy - bz \end{aligned}$$

See Chapter 3.

Measure

In mathematics a measure is a real number associated with a set (and having certain other properties). Perhaps the most familiar example is probability which is associated with subsets forming part of an event space. Most introductory texts gloss over the fact that the ordinary methods of calculus (Riemann integrals, for example) are inadequate for probability theory. Measure theory becomes even more important in dealing with fractal sets (see multifractal).

A18

Mode Locking

See phase locking, oscillator, circle map.

Multifractal

Fractals are sets having the property of self-similarity: multifractals are measures associated with sets. "Multifractal measures are related to the study of a distribution of physical or other quantities on a geometric support. The support may be an ordinary plane, the surface of a sphere or a volume, or it could itself be a fractal" (Feder, p.66). Multifractals arise when a measure defined on the support has different fractal dimensions on different parts of the support, in other words, where the measure shows spatial correlations. Such phenomena cannot be adequately described by a single fractal dimension, but require a complete spectrum of "dimensions". The concept was first developed by Mandelbrot for turbulence, and has been applied to a wide range of mathematical and geophysical phenomena, including the properties of strange attractors and aggregation phenomena. The use of multifractals permits a much more complete description of phenomena than can be expressed by a single dimension, therefore their use can be expected to increase rapidly in the future. For example, the structure of atmospheric turbulence, and the pore spaces in sandstones are known to show fractal geometry, and it is clear that they can only be described adequately by a model that includes spatial correlations. For an introduction and key references see Feder (1988) and Stanley and Meakin (1988).

A simple example of a multifractal is diffusion limited aggregation (DLA: Stanley and Meakin, 1988). In DLA, a "dendrite" grows by aggregation of particles onto a seed: the particles follow a random path, so that at a given stage of dendrite growth one can specify a probability, p_i, that the ith site on its surface will be hit by the next particle. The probabilities of all the sites constitute a probability distribution n(p), which can be characterized by its moments, Z(q)

$$Z(q) = \Sigma\, n(p).p^q$$

One can relate the moments to a characteristic diameter of the aggregate, L, by an exponential equation

$$Z(q) \sim L^{-\tau(q)}$$

It is generally assumed that only a small number of exponents are necessary, in other words the τ reduce essentially to a few "dimensions" (which are probably fractal). It has now been shown, however, that this is not true of DLA, and an infinite number of exponents are required. Generally investigators represent these, not as the function $\tau(q)$ itself, but as its Legendre transform $f(\alpha)$. It has been found that this function is a very useful way to characterize multifractals. It is smooth and continuous and shows a single maximum. It is also universal, in that the same general type of function characterizes many different types of multifractal phenomena, or an identical function characterizes a whole range of phenomena. For example, a single $f(\alpha)$ distribution characterizes fluid turbulence formed in many different ways at many different scales (see Feder, 1988, p.76-78).

Another example is a multifractal constructed on a Cantor set. Instead of the usual (one scale) construction, we introduce two scales, e.g. $s_1 = 1/4$ and $s_2 = 2/5$. The unit line is first divided into a segment from 0 to 1/4, and a second segment from 3/5 to 1 (i.e., the first segment has length 1/4, the second has length 2/5). A probability of $p_1 = 3/5$ is assigned to the first segment, and of $p_2 = 2/5$ to the second segment (i.e., the first segment has a larger "weight" than the second). The process is then repeated, to give four segments with length s_1^2, s_1s_2, s_2s_1, and s_2^2, and with probabilities p_1^2, p_1p_2, etc., and so on indefinitely. In this example, therefore, the support is a fractal set, the measure is a probability, and the spatial correlation is introduced by defining the probability in a way which depends upon position in the fractal set. The specified fractal set is then partitioned into subsets of length l_i, and to each subset it is also possible to assign a probability p_i. We assume that, in the limit as the subset length tends to zero, the probability is given by a power law

$$p = k\, l^{\alpha}$$

α is then a (fractal) dimension defined for a particular item in the (fractal) set, and there is a function $f(\alpha)$ which shows how this dimension is distributed over the whole set. The process can be repeated for all items in the set, and the results is a summed distribution of $f(\alpha)$. Such a distribution can be calculated, or derived by computer simulation.

The term "multifractal" can be related either to the observation that a small number of dimensions is not adequate to describe multifractal phenoemena or to the derivation of the measure (e.g., p in the Cantor example) from a union of many sets, each of which is fractal with its own dimension.

Noise

It is useful to divide a time series into the signal (the "desired" or "meaningful" part) and the noise (the "undesired" or "meaningless" part). The distinction is subjective: one person's noise is another person's music.

A common mathematical model for noise is given by:

$$y(t) = x(t) + \varepsilon$$

A20

where y(t) is the total time series, x(t) is the signal, and ε is the noise, assumed to be a random variable.

Certain types of noise may be defined precisely. Examples are white noise which consists of a uniform random distribution. Another is Gaussian noise, which is random (and so can be considered a variety of white noise), but has a Normal distribution.

Noise may be described by its appeared in a variance (power) spectrum, which plots the variance V(f) against the frequency f, for a range of frequencies observed in the observed time series. V(f) is also called the power or spectral density. White noise shows no systematic variation in variance with frequency. Much of the noise displayed by real spectra, however, shows that V(f) varies as $1/f^n$.

If noise is generated by summing (integrating) white noise, as in a random-walk, then it is described as Brownian, or Brown noise. This noise shows a spectrum where V(f) is proportional to $1/f^2$, that is, n = 2.

If we consider a time series composed only of random noise, then it is impossible to predict the next interval in the series from a knowledge of the series up to that time. It is still possible to state the probability that, in some time interval δt, the value of y will be less than some value y_o. Assume this probability is proportional to δt^H. For Brownian noise H = 0.5.

A common type of noise, whose origin is not yet fully explained is 1/f noise, which shows a spectrum with V(f) proportional to 1/f, that is n = 1.

A significant feature of $1/f^n$ noise is that it implies that there is no scale that adequately characterizes the noise: all possible scales are present, though the larger scales are proportionately less frequent (see scale, characteristic).

The fractal dimension is given by

$$n = 5 - 2D$$

For example, Brownian noise is a fractal with dimension 1.5.

Suppose that a time series relates the magnitude of an event (e.g., a flood or an earthquake) to the time at which it occurs. Then, for a given length of record the frequency can be plotted against the magnitude: such plots often show a power law relationship, i.e. they plot as straight lines on log-log graph paper. Such observations imply a spectrum with V(f) varying as $1/f^n$.

A21

Oscillator

A system displaying oscillatory motion. The simplest oscillator is a mass whose motion is controlled by an ideal spring. Its equation of motion is given by

$$d^2x/dt^2 = -w^2x$$

that is, the acceleration is proportional to the displacement (w^2 is the spring constant). A general form of the equation of a nonlinear, damped oscillator is

$$d^2x/dt^2 = k_1f_1(x) + k_2f_2(x)$$

where the k's are coefficients, and the f(x) are generally nonlinear functions. For a damped pendulum, for example, k_1 is $-g/L$, $f_1(x)$ is sin x, k_2 is a (negative) damping coefficient, and $f_2(x)$ is dx/dt. Oscillators may be forced by adding another periodic driving force to the equation. An example is described by the van der Pol equation

$$d^2x/dt^2 = (\varepsilon - x^2)\ dx/dt - x$$

The equation was derived by making the friction coefficient depend on the displacement, so that it is positive for large displacements but negative for small ones. Negative friction corresponds to forcing and is not meaningful for ordinary mechanical oscillators, but the equation does describe certain types of electrical oscillators. The equation has a limit cycle.

After an initial period of adjustment, driven oscillators may settle down to a regular oscillation where the periods of the two phases have been adjusted so that their ratio is rational: this is the phenomenon of phase locking.

Percolation Theory

The term was introduced by Broadbent and Hammersley (1957) for a new science dealing with clustering of particles by random processes. It includes the following problem: let particles occupy sites on a regular lattice (e.g., a two-dimensional square lattice), and define a cluster as a group of particles with nearest neighbour links to other particles in the cluster (that is, each particle connects above, below, or to the side with some other particle in the cluster). Assign a probability at random to each site, and then define whether a site is occupied or not by specifying a minimum probability, p, that any site is occupied by a particle. What must this probability level be so that there is a very high chance that a cluster spans the entire lattice (that is, it extends without a break from one side to another)? This can be defined as the percolation threshold for that lattice: if the lattice defines an idealized porous material, the "particles" are pore spaces, and the "clusters" are interconnected groups of pore spaces, then the percolation threshold would be the condition where a fluid could percolate through the porous material. It turns out that the transition from "non-percolating" to "percolating" lattices is generally quite abrupt (corresponding to a "critical" condition) and that, close to the critical condition, the sizes of clusters have interesting statistical properties: all possible sizes of cluster are present, and have the power law size distribution typical of fractals. Percolation theory applies not only to porous media but to aggregation of particles by random walk processes (thus to the formation of dendrites). It has much in common with the theory of phase transitions such as condensation from a vapour to a liquid or magnetization of minerals at the Curie Point (see renormalization group).

Phase Locking

"...a resonant response occurring in systems of coupled ocsillators or oscillators coupled to periodic external forces. In general, resonances occur whenever the frequency of a harmonic, $P\Omega_1$, of one oscillator approaches some harmonic, $Q\Omega_2$, of another; and in the resonant region the frequencies of the two oscillators lock exactly into the rational ratio P/Q" (Jensen et al., 1984, p.1960). see circle map and oscillator.

Phase Space

An ordinary differential equation of order n may be represented as a system of n first order ordinary differential equations. For example, the motion (in the x-direction) of a mass m suspended from a spring and subject to viscous damping (friction) can be described by the equation:

$$d^2x/dt^2 + r(dx/dt) + w^2x = -g$$

(see also oscillator). By letting y = x and z = dy/dt it is possible to write this equation as the system:

$$dy/dt = z$$

$$dz/dt = - w^2y - rz + g$$

The set of all possible solutions of such a system is sometimes called a flow in the space of two dimensions. Its behaviour may be studied by plotting trajectories from some arbitrary starting point, using y and z as the coordinates of the space, rather than the original x and t. The variables y and z are called state variables, because they define the possible states of the system. In the example, y is the position, and z is the velocity. The space with these variables is called the phase space, and a section through it is called a

Poincaré section. Strictly, a phase space is a plot of position vs velocity (or momentum). The term is also used loosely for any space, defined by a set of state variables, related by first order differential equations. If the variables are not those of position and velocity, such a space is more properly called a state space.

Poincaré Section

The behaviour of a system of ordinary differential equations may be described by constructing trajectories in phase (or state) space. It is not always easy to describe the complete trajectory, and therefore it is better in some cases to study the intersection of the trajectories with a single plane.

Commonly the position of a point on the Poincaré section is determined numerically by a process of numerical integration, i.e, it is calculated from the position of the previous intersection by solving the differential equations for the trajectory. The process can be called a mapping (or iterative map) of the points and represented symbolically:

$$P_{k+1} = T(P_k)$$

Note, however, a difference between a Poincaré section and a typical iterative map. The map is generally calculated by incrementing time repeatedly by a constant amount. The time taken for trajectories to cross the same plane in phase space, however, is not necessarily the same for succesive trajectories.

One graphic representation of the chaotic nature of dynamic systems is the scatter of points, and the unpredictable way that they appear on Poincaré sections, or on iterative maps. Some of the general laws governing the distribution of chaotic and regular periodic trajectories in phase space are given by the KAM theorem.

Probability (see also Random)

Over the years, several different definitions have been suggested for the concept of probability. The most common approach at present is to define probability axiomatically, but the classical definition is also used.

The classical approach is to consider an experiment (or trial), such as tossing a coin, or throwing a die, which may have a limited number of outcomes (head, tail; 1,2,3,4,5,6). We may further define the outcomes as being equally likely (by inference, or more realistically, by definition). A particular outcome is called an event (see random). Then if a trial may result in any one of N exhaustive, mutually exclusive, and equally likely cases, and if M of these produce the event, then the probability is M/N. Probabilities vary from zero (impossible) to one (certain), and follow certain rules. Several problems with this classical approach will occur to the reader: it depends upon the prior definition of "equally likely outcomes", which is itself a concept involving probability, and there is the problem of how the concept can be applied if the number of outcomes is infinitely large (i.e., not enumerable).

The axiomatic theory is based on a set S of elementary events and a system B of subsets of S. The elements of B (subsets of S) are called random events, and B is called a Borel field if it satisfies the following conditions:

(i) S is an element of B;
(ii) If two sets E1 and E2 are elements of B then their union, their intersection, and their complements are also elements of B;
(iii) If the sets E1, E2,... are elements of B then their union and their intersection are also elements of B.

Then Kolmogorov's system of axioms is:
(i) To every random event in the field of events there is assigned a non-negative real number called the probability;
(ii) The probability of the certain event is 1;
(iii) If the events E1, E2,... are mutually exclusive in pairs, then the probability of their intersection is equal to the sum of their individual probabilities.

Random

This simple term proves to be a hard one to define. If we conduct an experiment, which may be something as simple as flipping a coin, or drawing a ball from an urn containing a mix of black and white balls, and if we observe that the result of the experiment is not completely reproducible, and not completely predictable from a knowledge of previous results of running the same experiment, then we may say that the result is at least partly produced by random events.

If it is possible to enumerate all possible outcomes of a random experiment, then this set can be called the sample space. For example, in tossing a coin we recognize heads and tails as the two possible outcomes. Then an event is a subset of a sample space. For example, if we toss the coin once, the sample space consists of one head and one tail, and the event will be one or the other (this is an elementary event). If we toss the coin twice in succession the sample space would be: HH, HT, TT, TH. An event would then be one of these four possibilities. Instead of using words or letters we could assign a number to each outcome. Then we could define a random variable as a function connecting a set of outcomes to the experiment to the set of numbers that correspond to these outcomes. A little more formally we say that a random variable f(z) describable by a sample point, z, is a function defined on the sample space of an experiment such that for every real number of the variate, z, there exists a probability p(z).

A25

We see that the concept "random" is closely connected with the concept of sampling. Sampling implies the (at least potential) existence of a set of possible events, which is the object of investigation, and the actual subset of observed events. The former is the population, and the latter is the sample.

A sample may be defined as a random sample if every item in the population has an equal chance of being chosen. This further implies that the result of one "choice" (trial, draw, experiment) does not influence in any way the result of the next one. Alternatively, a random sample may be defined as a set of independently and identically distributed random variables.

An interesting critique of the distinction generally drawn between "random" or "stochastic" processes, and deterministic processes has been given by Ford (1983). One problem is that some fully deterministic processes give rise to outcomes which, to an observer who does not know the generating process, appear to be completely unpredictable and therefore "random". In fact, since a coin toss follows simple dynamic laws, we might say that it is exactly such a process. Interestingly enough, the pioneers of statistics actually conducted long coin-toss experiments, which "verified" experimentally some of the fundamental theorems of probability theory (e.g., the law of large numbers). However, such theorems can no more be verified by experiment than the theorems of geometry can be. All that can be verified is that, under certain experimental conditions, the axioms of probability theory, and deductions from them, correspond reasonably closely to observable phenomena.

Renormalization Group

A scaling technique developed originally in quantum theory and applied by K.G. Wilson to explain phase transitions (see Wilson, 1979). "Renormalization" means the repeated rescaling of variables so that the general form of the equations connecting them is preserved. The variables constitute a "group" in the mathematical sense of a set whose elements can be combined together (in this case rescaled) in a way having definite mathematical properties. Application of the technique reveals that close to phase transitions, physical variables can be related by power equations, and that there are certain values for the exponents in these equations that remain invariant under repeated rescaling: these values define the "fixed points" of the system, and may have very general significance, which is independant of the details of the theoretical model used to derive them.

Renormalization-group methods have been applied to percolation phenomena, many of which have fractal properties. The self-similarity of fractals makes them a natural field of application of these methods. Simple introductions are given by Wilson (1979), Feder (1988), Gould and Tobochnik (1988, v.2), Stauffer (1985), and Zallen (1983). For a historical summary see Nelson (1981).

A26

Return Map

Given a map $x_{k+1} = f(x_k)$, a first return map plots values of x_k against x_{k+1}. It is also possible to plot return maps for higher iterations (e.g., k and k+2). Complicated strange attractors may be easier to visualize by displaying two dimensional sections through them. But an even more striking simplicity is often seen by plotting the first return map. As this simplicity is not expected from a truly random map, its existence is evidence for nonlinear dynamical chaos. The simplicity of the return map also allows it to be characterized by a numerically fitted function, say g. Then it is possible to determine the largest Lyapunov exponent of the original system from this function. See Bergé et al. (1986, Appendix B) for details.

Rössler attractor (or equations)

A particularly simple set of equations with a strange attractor, first proposed in 1976. The equations are

$$dx/dt = -(y+z)$$
$$dy/dt = x + ay$$
$$dz/dt = b + z(x-c)$$

It has been found that a similar attractor is displayed by a some real systems. A now classic example is a dripping tap (Crutchfield et al., 1986).

Scale, Characteristic

Individual phenomena generally may be partly described by a scale; for example, a rock fragment may be characterized by its volume or diameter, and an earthquake by its Richter magnitude. An observed sample of such phenomena shows a range of such scales, but the variation can be described by statistical measures such as the mean and standard deviation. It is often assumed that the entire natural population also can be characterized by such scales: this would be correct, for example if the size of rock fragments was Normally (or Lognormally) distributed. But if the size follows a power law distribution, that is, if it shows a fractal distribution, then there is no characteristic physical scale. There is no meaningful "average size" of a true fractal size population.

The notion that there is no characteristic scale to a group of phenomena is counterintuitive, because our experience indicates that even if a part of the distribution is fractal, in nature there must be upper and lower limits. For, example rock fragments cannot be smaller than the unit cell of mineral grains making up the rock, or larger than the entire earth. But observation indicates that many phenomena are fractal over a very large range.

The fact that fractal distributions have no characteristic scale suggested to Bak et al. (1987) that they indicate a system maintained in a critical state. It is known that critical transitions, such as phase transitions, the Curie point in magnetism, or the critical state in percolation theory, are characterized by the presence of all possible scales of phenomena (e.g., all possible sizes of ordered domains or interconnected clusters of pore spaces). Bak has suggested, therefore, that if all possible scales are continuously present, the system is probably in a critical state, and must be maintained in this state by the operation of the system itself. In this sense, the system shows evidence of "self-organized criticality".

Self-Affinity

When a geometric object, for example, a cube is scaled to a different size, each of its dimensions is scaled by the same factor. Such a scaling can be described as due to a similarity transformation. Objects which are strictly similar to each other can be transformed into each other by such a transformation (see self-similarity).

Under an affine transformation, different dimensions are scaled by different factors. If a part of an object can be transformed into a smaller (or larger) part of the same object using an affine transformation the object is described as described as self-affine. Just as with similarity, affinity may be "absolute" or it may be statistical. In statistical affinity the parameters describing the two statistical distributions undergo an affine transformation.

Self-Organization

In the context of chemical systems, this term was popularized by Nicolis and Prigogine (1977). Strangely, though the term appears in the title of their books, it is nowhere given a succinct definition. Perhaps their closest approach is the statement (p.5):

> "...self-organization in nonequilibrium systems [is] characterized by the appearance of dissipative structures through the amplification of appropriate fluctuations."

Paradigms include purely physical systems, such as thermal convection in fluids, and oscillating chemical systems such as the Belousov-Zhabotinski reaction, and Liesegang phenomena.

Ortoleva et al. (1987) have given a general discussion of self-organization in geochemical reaction-transport systems, and give the following definition (p.980):

> "Self-organization is the autonomous passage of a system from an unpatterned to a patterned state without the intervention of an external template."

They point out two necessary conditions: that the system be far from equilibrium (therefore controled by nonlinear processes) and that "at least two processes active in the system be coupled" (p.980).

The study of self-organization therefore overlaps to a large extent with the study of nonlinear dynamics, though it concentrates on the appearence of pattern, rather than on chaotic phenomena. See also Brusselator.

Self-Similarity

This is a key concept in fractals. Self-similarity means that a part of the whole shows the essential features of the whole. In the case of fractal curves, such as the Koch Curve the similarity may be exact, in the sense that a small part of the curve may be magnified to reproduce exactly the features of a larger part of the curve. The self-similarity of many natural fractals, however, is statistical rather than exact, that is, small parts of the system shows the same statistical properties as larger parts. This implies that the system has no characteristic physical scale (see also self-affinity).

"...the concept of self-similarity is certainly not new...It has been well known since the detailed study of critical properties of phase transitions...and has been instrumental in the formulation of the renomalization group approach...that has essentially solved this problem. In this case, however, self-similarity was considered a peculiarity of the competition between order and disorder at a particular temperature for equilibrium phase transitions. We can now see that this property is much more common and appears in many equilibrium and nonequilibrium phenomena apparently unrelated to the problem of phase transitions." (Pietronero, 1988, p.278).

Sierpinski Gasket and Carpet

The Sierpinski gasket is a fractal produced by starting with a filled triangle, cutting out a central triangle (with a side one half of the original), and then repeating the process indefinitely. The fractal dimension is 1.58... The carpet is similar except that one begins with a square, and cuts out a central square with a side one third of the original. The fractal dimension is 1.89... These and other Sierpinski figures can also be generated in other ways.

State Space

See phase space.

A29

Stochastic

The word is derived from the Greek, meaning to contemplate or conjecture, and its use in mathematics can be traced back to Jacob Bernoulli. The term is often used as a synonym of "random", but generally as applied to a "process" (e.g., stochastic process, random process). Such a process is generated by one or more random variables (see random) which in turn depend on certain space and/or time parameters. For example, a process that gives rise to a "random walk" is a stochastic process. The outcome of a stochastic process, therefore, cannot be exactly predicted: at best, it is possible to associate probabilities with the range of possible outcomes. Most stochastic processes attempt to combine purely deterministic phenomena with a degree of random behaviour. For example, a stochastic theory of ballistics might combine the usual deterministic theory based on Newtonian mechanics, with estimates of the probable error in prediction of such a theory based upon the addition to the theory of certain random variables.

Symbols

Much of the literature of Nonlinear Dynamics makes use of concepts and symbols of branches of pure mathematics that are not well-known to many scientists. Useful reference works that explain these concepts include Boroski and Borwein (1989), Daintith and Nelson (1989) and Jackson (1989).

The following is a short list of the most commonly used notations.

Sets

A set consists of a collection of defined elements, e.g., the set of real numbers $\mathbb{R}$, or a set of points in a two-dimensional Euclidian space $\mathbb{R}^2$. The set of elements x, having a defined property P(x) may be written

$$\{x|P(x)\}$$

for example, if x lies in the range $0<x<1$, we write

$$\{x|0<x<1\}$$

That x is an element of the set X is written $x \in X$.

$X \subset Y$ means that set X is a subset of set Y.

$X \cup Y$ is the union of sets X and Y, i.e. the set consisting of all the elements of both X and Y.

A30

$X \cap Y$ is the intersection of sets X and Y, i.e. the set consisting of all the elements that are common to both X and Y.

The Cartesian product of two sets X and Y is written $X \times Y$ and consists of the set of all ordered pairs (x,y). This definition can be written

$$X \times Y = \{(x,y):(x \in X) \;\&\; (y \in Y)\}$$

Functions

The familiar notation

$$f(x) = \mu x(1-x)$$

defines a function, f, of x. We generally think of functions as continuous, in the sense that they can be represented (in two dimensions) by a smooth line. A continuous function is abbreviated C^n, where n refers to the highest order derivative that the function possessess at every point in the domain of interest.

Maps

The concept of a function is frequently extended to discrete maps, such as

$$x_{k+1} = \mu x_k(1-x_k)$$

which defines the value of x for the (k+1)th step, in terms of its value at the kth step. A map associates each element x of one set X with an element y in a second set Y. This can be written

$$f:X \rightarrow Y \quad \text{or} \quad f|X \rightarrow Y$$

or

$$f:x \mapsto y$$

In words, $x \mapsto y$ means "x maps to y".

Composition

Repeated application of a discrete mapping can be represented

$$f^n(x) = f \circ f \circ \ldots f$$

Note that this does not mean f raised to the nth power, but f applied n times to some starting value of x.

A31

Torus

A doughnut-shaped surface. The ideal torus is formed by revolving a circle about an axis which is in the plane of the circle but does not intersect the circle. The term may be used by extension to any surface in phase space, whose Poincaré section lies on a closed loop. Trajectories in the three-dimensional phase space wind around the surface of the torus, which constitutes a limit cycle. Such trajectories are characterized by two frequencies, one related to motion around the torus itself (as seen in a "plan view" or axial projection) and the other to motion around sections through the torus. The winding number is the ratio of these two frequencies: if it is rational, then the trajectories repeat themselves after a number of paths around the torus and the Poincaré section consists of a number of discrete points; if it is irrational, then the trajectories cover the entire surface of the torus, and the Poincaré section consists of a continuous line that forms a loop.

Van de Pol equation

A second order differential equation with a limit cycle. See oscillator.

Winding Number

see circle map.

Acknowledgments

I wish to acknowledge corrections to an earlier draft by David Goodings.

References

Arnol'd, V.I., 1986, Catastrophe Theory. N.Y., Springer-Verlag, Second Edition, 108 p.

Bak, Per, 1986, The devil's staircase. Physics Today, December 1986, p.38-45.

Bak, Per, C. Tang and K. Wiesenfeld, 1987, Self-organized criticality: an explanation of 1/f noise. Physical Review Letters, v.59, p.381-384.

Bergé, Pierre, Y. Pomeau, and C. Vidal, 1986, Order within Chaos. N.Y., John Wiley, 329 p.

A32

Boroski, E.J., and J.M. Borwein, 1989, Dictionary of Mathematics. London, Collins, 659 p.

Broadbent, S.R., and J.M. Hammerseley, 1957, Percolation processes I. Crystals and mazes. Proc. Cambridge Philosophical Soc., v.53, p.629-641.

Crutchfield, J.P., J.D. Farmer, N.H. Packard, and R.S. Shaw, 1986, Chaos. Scientific American, v.255, no. 6, p.46-57.

Daintith, John and R.D. Nelson, eds., 1989, The Penguin Dictionary of Mathematics. London, Penguin Books, 350 p.

Essex, Christopher and M.A.H. Nerenberg, 1990, Fractal dimension: limit capacity or Hausdorff dimension? Amer. Jour. Physics, v.58, p.986-988.

Holden, Arun V., ed., 1986, Chaos. Princeton Univ. Press, 324 p.

Feder, Jens, 1988, Fractals. N.Y., Plenum Press, 283 p.

Ford, Joseph, 1983, How random is a coin toss? Physics Today, April 1983, p.40-47.

Glass, Leon, and M.C. Mackey, 1988, From Clocks to Chaos: The Rythms of Life. Princeton Univ. Press, 248 p.

Gould, Harvey and Jan Tobochnik, 1988, An Introduction to Computer simulation Methods: Applications to Physical Systems. Reading, MA, Addison-Wesley Publ. Co., 2 volumes.

Hénon, Michel, 1969, Numerical study of quadratic area-preserving mappings. Quarterly of Applied Math., v.27, p.291-311.

Hénon, Michel, 1976, A two-dimensional mapping with a strange attractor. Communications in Mathematical Physics, v.50, p.69-77.

Hénon, Michel, and Carl Heiles, 1964, The applicability of the third integral of motion: some numerical experiments. Astronomical Jour., v.69, p.73-79.

Holden, Arun V., ed., 1986, Chaos. Princeton, NJ, Princeton University Press, 324 p.

Jackson, E. Atlee, 1989, Perspectives of Nonlinear Dynamics, Volume 1. Cambridge, Cambridge University Press, 495 p. (See Appendix A and B, p.376-392).

Jensen, M.G., P. Bak, and T. Bohr, 1984, Transition to chaos by interaction of resonances in dissipative systems. I. Circle maps. Physical Review, v.A30, p.1960-1969.

Li, T.Y. and J.A. Yorke, 1975, Period three implies chaos. Amer. Mathematical Monthly, v.82, p.985.

Lichtenberg, A.J., and M.A. Lieberman, 1983, Regular and Stochastic Motion. N.Y., Springer-Verlag, 499 p.

Mandelbrot, Benoit B., 1982, The Fractal Geometry of Nature. N.Y., W.H. Freeman and Co., 468 p.

Nelson, David R., 1981, Renormalization. in Rita G. Lerner and George L. Trigg, eds., Encyclopedia of Physics. Reading, MA, Addison-Wesley Publ. Co., p.876-878.

Nicolis, G., and I Prigogine, 1977, Self-Organization in Nonequilibrium Systems. From Dissipative Systems to Order through Fluctuations. N.Y., John Wiley and Sons, 491 p.

Ortoleva, P., E. Merino, C. Moore, and J. Chadham, 1987, Geochemical self-organization I: reaction-transport feedbacks and modeling approach. American Jour. Sci., v.287, p.979-1007.

Percival, Ian and Derek Richards, 1982, Introduction to Dynamics. Cambridge, Cambridge Uniiversity Press, 228 p.

Pietronero, L., 1988, Fractals in physics: introductory concepts. in S. Lundqvist, N.H. March and M.P. Tosi, eds., Order and Chaos in Nonlinear Physical Systems. N.Y., Plenum Press, p.277-294.

Prigogine, Ilya, 1980, From Being to Becoming: Time and Complexity in the Physical Sciences. N.Y., W.H. Freeman and Co., 272 p.

Ruelle, D., and F. Takens, 1971, On the nature of turbulence. Communications in Math. Physics, v.23, p.343.

Ruelle, D., 1980, Strange attractors. Math. Intelligencer, v.2, p.126.

Sander, Leonard M., 1987, Fractal growth. Scientific American, v.256, no. 1, p.94-100.

Smale, S., 1967, differentiable dynamical systems. Bull. Amer. Math, Soc., v.13, p.747.

A34

Sorensen, Peter R., 1984, Fractals. Byte, Sept. 1984, p.157-172.

Stanley, H. Eugene, and Paul Meakin, 1988, Multifractal phenomena in physics and chemistry. Nature, v.335, p.405-409.

Stauffer, Dietrich, 1985, Introduction to Percolation Theory. Philadelphia, Taylor and Francis, 124 p.

Thompson, J.M.T., and H.B. Stewart, 1986, Nonlinear Dynamics and Chaos. N.Y., John Wiley and Sons, 376 p.

Tietjen, Gary L., 1986, A Topical Dictionary of Statistics. N.Y., Chapman and Hall, 171 p.

Walker, Grayson H., and Joseph Ford, 1969, Amplitude instability and ergodic behavior for conservative nonlinear oscillator systems. Physical Review, v.188, p.416-432.

Wilson, Kenneth G., 1979, Problems in physics with many scales of length. Scientific American, v.241, no.2, p.158-179.

Zallen, Richard, 1983, The Physics of Amorphous Solids. N.Y., Wiley-Interscience, 304 p.

Zeeman, E.C., 1976, Catastrophe theory. Scientific American, v.234, no.4 (April), p.65-83.

B1

APPENDIX 2: COMPUTER PROGRAMS

Gerard Middleton
Department of Geology
McMaster University
Hamilton ON
L8S 4M1

A.1 Introduction

Inside the back cover of this volume there is a single double-sided, double-density 5 1/4" floppy disk, formatted by MS-DOS, and suitable for running in most IBM-compatible computers. It contains a number of programs written in Turbo Pascal (Copyright Borland International, 1984, 1989). Certain support files are also included, notably the .BGI graphics files, which remain the property of Borland International and may not be sold or distributed separately.

The programs will run best on computers equipped with EGA or VGA color graphics. It is not necessary to have a hard disk or a math coprocessor. The programs make use of colour but should also run on machines with a monochrome Hercules or VGA card. They should also run, imperfectly on machine equipped only with CGA.

Colour is used not only to distinguish different trajectories on the same graph, but also in the logistic map to indicate superimposition of plotted points.

The programs are designed for educational use only, and are not guaranteed in any way. The writer retains the copyright, but authorizes rewriting and distribution of single copies of the programs, provided that the source is acknowledged and the programs are not sold for profit.

To start the programs, put the disk in Drive A: of your computer, change the default drive to drive A:, and either type README for more information, or type CHAOS to begin the first program. The disk with the .BGI files MUST be in the default drive: so if you intend to run these programs from a hard disk be sure to copy these files to your disk, along with the .EXE programs.

If you require a paper copy of the graphs, use a graphics print-screen utility suitable for your own computer.

The programs on the disk include:

:CHAOS
:CRVSR

The CHAOS program includes several subprograms ("Procedures" in Pascal jargon) which are offered by the following opening menu:

Choose from the following menu:

1. Logistic map;
2. Oscillator and forced Van der Pol oscillator;
3. Lorenz attractor;
4. Rössler attractor;
5. Hénon quadratic map;
6. Hénon attractor;
7. Circle map;
8. Quit.

A guide to the use of these programs follows.

A.2 Logistic Map

This program demonstrates the logistic map, which is described in Chapter 5, and in most introductory texts on chaos.

Try running the program first with r in the range from 3 to 4 (using an arithmetic scale), and with x in the full range from 0 to 1. Next try r on a logarithmic scale in the range from 3 to 3.5699: this demonstrates the doubling of bifurcations, with the increment in r between bifurcations decreasing in a roughly geometric progression (equal spacing on the log scale).

Colour is used to indicate the superimposition of plotted points. Each time a new point is plotted in the same position (pixel) as a previously plotted point, the colour changes in the following sequence:

B3

1. Blue, 2. Green, 3. Cyan, 4. Red, 5. Magenta, 6. Brown, 7. Light Gray, 8. Dark Gray
9. Light Blue, 10. Light Green, 11. Light Cyan, 12. Light Red, 13. Light Magenta
14. Yellow, 15. White.

A.3 Oscillator and forced Van der Pol oscillator

This program is designed to introduce the concepts of phase space, trajectories in such spaces, and attractors. The example used is the oscillator. The introductory screen gives the equation for a typical oscillator:

$$\frac{d^2\theta}{dt^2} = k_1 f_1(\theta) + k_2 f_2(\theta)$$

θ is an angular measure of displacement (the angle from the vertical, in the case of a pendulum). It is measured in radians and varies from -pi to +pi. k_1 and k_2 are numerical coefficients, such as damping coefficients, and f_1 and f_2 are functions of θ.

The oscillator equation is a second-order differential equation, so can be reduced to two first order equations by using the transformation $d\theta/dt = y$ as one of the equations. The phase space is two dimensional and consists of the θ - $d\theta/dt$ plane.

The program offers four options:

1: Simple Pendulum

$k_1 = -g/L$, $k_2 = 0$, $f_1 = \theta$

where g is the acceleration due to gravity, and L is the length of the pendulum. In the program k_1 is set equal to -1. Such a pendulum is an example of a linear Hamiltonian system. There is no friction, therefore no attractor. Trajectories in phase space are closed curves.

2: Nonlinear Pendulum

$k_1 = -g/L$, $k_2 = 0$, $f_1 = \sin\theta$

A nonlinear Hamiltonian. The trajectories are very similar to those of the linear pendulum, but differ somewhat in their shape.

B4

3: Damped Pendulum

$k_1 = -g/L$, $k_2 < 0$, $f_1 = \sin\theta$, $f_2 = d\theta/dt$

This system is dissipative, and has a point attractor at the origin. This corresponds to the pendulum hanging vertically downwards. Trajectories spiral in towards the attractor.

4: Van der Pol oscillator

$k_1 = -1$, $k_2 < 0$, $f_1 = \sin\theta$, $f_2 = (d\theta/dt)(\theta^2-1)$

This system is also dissipative ($k_2 < 0$) but also driven. After an initial "transient" it settles into a regular orbit around the origin. The regular orbit is a one-dimensional (line) attractor, called a limit cycle.

A good introductory discussion of oscillators is in Bergé et al. (1986, Chapter 1), Baker and Gollub (1990, Chapter 3) and Rasband (1990, Chapter 1 and 6). Driven damped oscillators frequently give rise to chaotic attractors of great complexity: an example is the Duffing equation (see Rasband, 1986, Chapter 6). As Rasband (1986, p.128) points out

> "The key point in the dynamics of the damped, driven pendulum is that the strong damping causes the Poincaré map...to reduce to a one dimensional map on the circle."

For these circle maps, see below.

B.4 Lorenz attractor

The Lorenz equations are discussed in Chapter 3. The program is designed to display the equations in the range $24 < r < 28$, though that range can be extended somewhat. There are other interesting ranges of r for the Lorenz equations, and the program can easily be adapted to them by adjusting the scale for the plots in the source code.

The program offers two options:

1: A plot of trajectory in the XZ or YZ plane of phase space;

2: A Poincaré section, for the plane Z = (r-1), and a first return map of Zmax(k) against Zmax(k+1).

B5

In both cases, X,Y, and Z are also plotted (in different colours) against time in a lower window.

The trajectory plots are straightforward, but are useful aids in visualizing the geometry of the attractor. Try both the XZ and YZ projections.

The Poincaré section is also straightforward. It appears to consist of two "lines" of dots, corresponding to a section through the two "wings" of the Lorenz "butterfly": the simplicit is misleading, since if it were enlarged sufficiently it would be possible to see that the cross-section of each "wing" consists of a narrow cloud of dots, rather than a true line. This is because the attractor consists not of a single plane, but of very many closely packed sheets.

The return map plots successive maximum values of Z (i.e., one dot is plotted each time the trajectory goes over the top of a loop). It is simpler than the Poincaré section, but contains within itself the essential properties of the Lorenz attractor, in much the same way that the return map of the Logistic map does (see the discussion in Chapter 5; and for more details see Bergé et al., 1986, Chapter VI).

B.5 Rössler attractor

This program uses a Runge-Kutta method to solve the equations: the method is fully described in Press et al. (1989).

The program is straightforward, and offers a choice of a perspective view of the trajectory in phase space, or an XY projection.

B.6 Hénon quadratic map

The program was modified from one published by Hughes (1986). It displays trajectories of the Hénon quadratic area-preserving map, and shows many of the features of such maps. See the entry KAM Theorem in the Chaossary.

B.7 Hénon attractor

This is a classic two dimensional mapping with a strange attractor. The program makes it possible to zoom in parts of the attractor, to see the increasingly detailed structure visible at high magnifications. The attractor can be seen to be fractal, in the general sense that it is made of parts similar in some way to the whole. See Bergé et al. (1986, p.130-135) for further discussion.

B6

B.8 Circle map

The particular map displayed is the two-dimensional sine circle map. Further discussion is provided in Chapter 8, and by Procaccia (1988) and Gunaratne et al. (1988).

The equations of the map are:

$$\theta_{n+1} = \theta_n + \Omega - (K/2\pi)\sin 2\pi\,\theta_n + Br_n \qquad \text{mod } 1$$

$$r_{n+1} = Br_n - (K/2\pi)\sin 2\pi\,\theta_n$$

These equation describe the motion of two oscillators. The frequencies are different, and in the ratio Ω. Each oscillator will move indepently of the other unless they are coupled together in some way. K measures the strength of the coupling. B is a factor that controls the strength of the radial distortion: for B = 0, the orbits are circles (the true, or simple, sine circle map) for B > 0 the radius or the "circle" varies.

The simple circle map has the equation:

$$\theta_{n+1} = \theta_n + \Omega - (K/2\pi)\ \sin 2\pi\,\theta_n \qquad \text{mod1}$$

The way the behaviour varies with Ω and K is shown in Figure B.1. For K = 0 the frequencies of the two oscillators are independent, and any ratio is possible. For K > 0, however, if the two frequencies are close to rational ratios (such as 1:1, 1:2, etc) the phenomenon of phase locking occurs, i.e., after an initial period, the frequencies adjust to the rational ratio, even if they were not exactly in that ratio to begin with. The regions in which this happens expand as K increases, as shown by the lines "splitting" and diverging toward the top of Figure B.1. These are the regions called "Arnol'd tongues".

To experiment with the circle map program, chose a value of B (start with B = 0), and a value of Ω, and observe what happens as the value of K is increased.

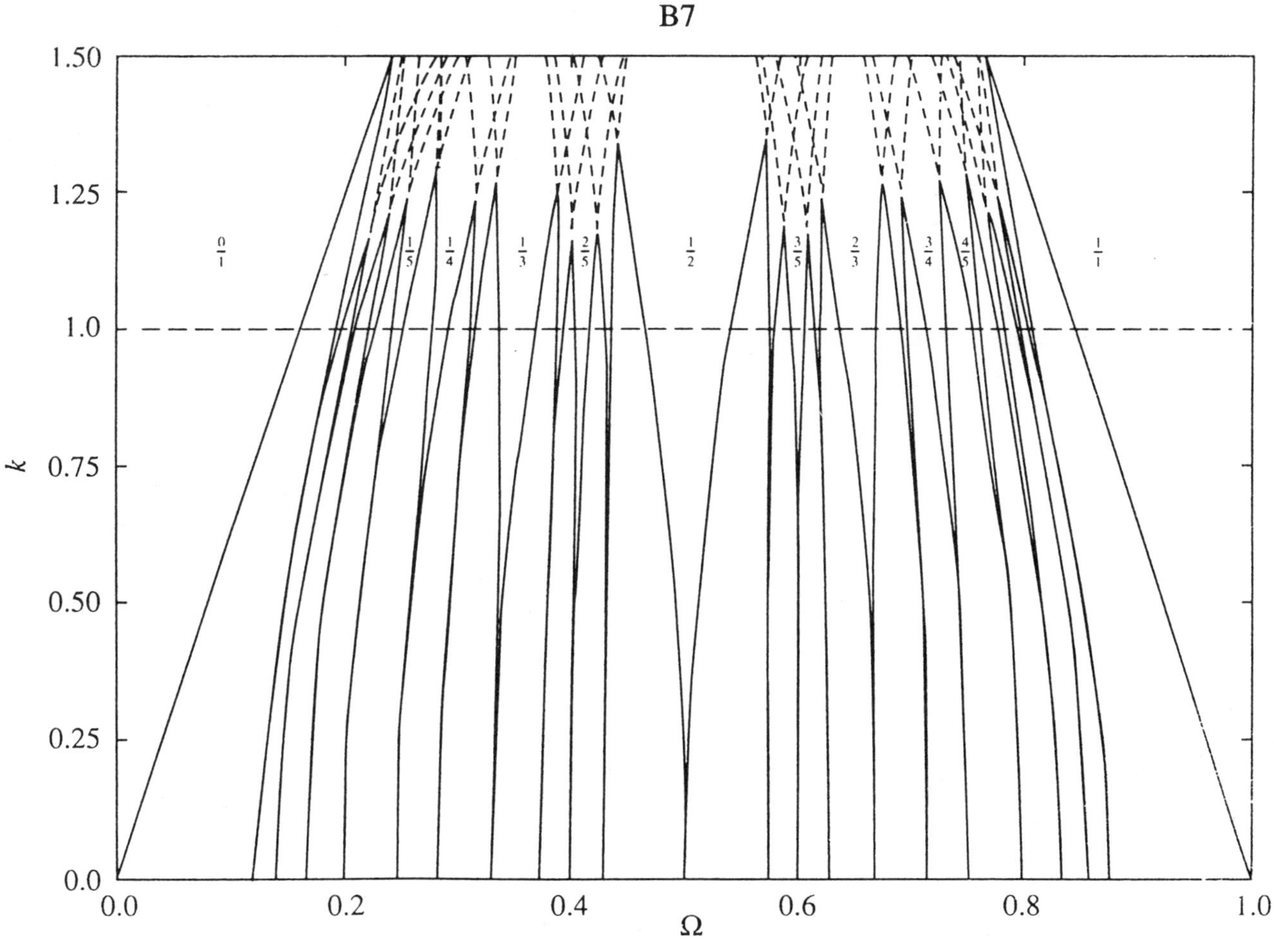

Figure B.1 Regions of phase locking in the simple circle map (Arnol'd tongues). From Jensen et al. (1984).

B.9 Crvsr

This program calculates the correlation dimension of a time series, by the method described in Chapter 4. The program is provided in a version suitable for demonstration, but the source code can easily be modified for research use.

The version provided acts on a time series of 430 data. This is generally not long enough to yield accurate results, but the number of arithmetic operations is very large: on an IBM AT, the program may run for more than 15 minutes, and on a fast 386 machine for more than 2 minutes. To process a time series with more than 2000 points, which is generally considered a minimum for accurate results in low dimensional systems (dimensions in the range 1 - 4) several hours might be required.

B8

The program offers a choice of operating on a random number sequence, generated by the computer itself, or reading in data from a Turbo Pascal real number file, produced by a program such as those demonstrating the Hénon attractor, the Lorenz attractor, or the Rössler attractor (see above). A sample is provided on the disk as HENON.DAT. The statistics and log-log plot are generated for embedding dimensions of 2, 3, 4, and 5.

Run the program first for random data, and observe that the slope of the plot continues to increase for each increase in embedding dimension. The number of data is not large enough to yield a good straight line plot, or an accurate estimate of the dimension. It is possible that the relatively poor results may also be partly due to a poor random number generator.

Next run the program for the Hénon data. It is observed that the slope is somewhat better defined, and remains almost constant, with a value of 1.2 to 1.3 in the central range. The known dimension of the Hénon attractor is 1.26. A file of data from the Rössler attractor is also provided. The data files may be examined using the Program ReadRfil.

B.10 Other published chaos programs

There are now several published sources of programs on chaos, written for educational purposes. Most do not provide source code. A few are in the public domain: perhaps the best of these is FRACTINT. The main purpose of this program is to generate fractal images based on the Mandelbrot and Julia sets, but the latest versions include several chaos programs too. The program (including source code in C) is available on many electronic bulletin boards, or by mail from Bert Tyler, Tyler Software, 124 Wooded Lane, Villanova PA 19085 (Tel. 215 525-6355).

Other sources are listed in the Bibliography.

Bibliography

Baker, Gergory L., and Jerry P. Gollub, 1990, Chaotic Dynamics, An Introduction. Cambridge, U.K., Cambridge Univ. Press, 182 p.

Bergé, Pierre, Y. Pomeau, and C. Vidal, 1986, Order within Chaos. Towards a deterministic approach to turbulence. N.Y., John Wiley and Sons, translated from the 1984 French edition by L. Tucherman, 329 p. (Still the best introduction).

Borrelli, Robert L. and Courtney S. Coleman, 1987, Differential Equations: A Modeling Approach. Englewood Cliffs, NJ, Prentice Hall Inc., 607 p. + 77 p. Appendix (A very useful introduction to nonlinear dynamics).

B9

Devaney, Robert L., 1990, Chaos, Fractals and Dynamics: Computer Experiments in Mathematics. Science TV, 181 p. + 2 videocassettes.

Gunaratne, Gemunu H., M.H. Jensen and I. Procaccia, 1988, Universal strange attractors on wrinkled tori. Nonlinearity, v.1, p.157-180.

Hughes, Gordon, 1986, Henon mapping with Pascal. Byte, v. 11, no. 13 (December 1986), p.161-178.

Jensen, M.H., P. Bak, and T. Bohr, 1984, Transition to chaos by interaction of resonances in dissipative systems. I. Circle maps. Physical Reviews A, v.30, p.1960-1969.

Koçak, Hüseyin, 1989, Differential and Difference Equations through Computer Experiments: with a supplementary diskette comtaining PHASER: An Animator/Simulator for Dynamical Systems for IBM Personal Computers. N.Y., Springer-Verlag, 224 p. (A useful set of programs and instructional text: VGA version now available.)

Parker, T.S., and L.O. Chua, 1990, Practical Numerical Algorithms for Chaotic Systems. N.Y., Springer-Verlag, 348 p. (Gives pseudocode for many algorithms).

Peitgen, Heinz-Otto, and Dietmar Saupe, eds., 1988, The Science of Fractal Images. N.Y., Springer-Verlag, 312 p. (Mainly on fractals, but includes a section by Devaney on Nonlinear dynamics: some chapters give pseudocode for algorithms).

Pickover, Clifford A., 1990, Computers, Pattern, Chaos and Beauty: Graphics from an Unseen World. N.Y., St. Martin's Press, 394 p. (Wide range of illustrations. Includes very sketchy pseudocode for some algorithms).

Press, William H., et al., 1989, Numerical Recipes in Pascal: the Art of Scientific Computing. Cambridge Univ. Press, 759 p.

Procaccia, Itamar, 1988, Universal properties of dynamically complex systems: the organization of chaos. Nature, v.333, 16 June, p.618-623.

Rasband, S. Neil, 1990, Chaotic Dynamics of Nonlinear Systems. N.Y., John Wiley and Sons, 230 p.

Index of Names

(does not include references in the Appendices)

Index of Subjects

(an asterisk indicates an entry in the Chaossary, Appendix 1: note that a brief guide to mathematical notation may be found under Symbols*)